Oxford Physics Series

General Editors

E. J. BURGE   **D. J. E. INGRAM**   **J. A. D. MATTHEW**

Oxford Physics Series

E. J. BURGE

Professor Emeritus, Royal Holloway and Bedford New College,
University of London

Atomic nuclei and their particles

Second edition

CLARENDON PRESS · OXFORD
1988

Oxford University Press, Walton Street, Oxford OX2 6DP

Oxford New York Toronto
Delhi Bombay Calcutta Madras Karachi
Petaling Jaya Singapore Hong Kong Tokyo
Nairobi Dar es Salaam Cape Town
Melbourne Auckland

and associated companies in
Beirut Berlin Ibadan Nicosia

Oxford is a trade mark of Oxford University Press

Published in the United States
by Oxford University Press, New York

British Library Cataloguing in Publication Data

Burge, E. J.
Atomic nuclei and their particles.—
2nd ed.—(Oxford physics series; no. 13).
1. Nuclear physics
I. Title
539.7   *QC*776
ISBN 0–19–851856–0
ISBN 0–19–851872–2 *Pbk*

Library of Congress Cataloguing in Publication Data

Burge, E. J. (Edward James)
Atomic nuclei and their particles.
(Oxford physics series; 13)
Bibliography: p. Includes index.
1. Nuclear physics. 2. Particle physics.
I. Title. II. Series.
*QC*776.*B*87   1988   539.7   87–31323
ISBN 0–19–851856–0
ISBN 0–19–851872–2 *Pbk*

Phototypeset by Macmillan India, Limited, Bangalore 25.
Printed and bound in
Great Britain by Biddles Ltd,
Guildford and King's Lynn

Editor's Foreword

Nuclear physics, and especially the particles of nuclear physics, have always provided a fascination for school students, undergraduates, and research workers alike. It is a subject which is not only of inherent interest but also forms part of the basic framework of any course in modern physics. It is for this reason that the Oxford Physics Series have included this as one of their core texts.

The text is written with a careful balance between ideas on the pure science side, such as those on symmetry and the conservation laws, and the implications on the applied side, as exemplified by fission and fusion. In a text of limited length many topics have necessarily received rather brief treatment, but readers will find that the topics treated reach to our current boundaries of knowledge, and reference forward to more detailed work is included.

D. J. E. I.

Preface to second edition

Advances in nuclear and particle physics, both theoretical and experimental, have been considerable since the first edition in 1977. They include such fundamental achievements as the unified theory of the electromagnetic and weak-nuclear interactions, confirmed by the evidence of the W and Z particles. The discovery of the tauon, the third lepton, implied top and bottom quarks and their associated characteristic particles. These experimental advances depended on outstanding developments in both accelerator and detector techniques, and involved teamwork between hundreds of physicists and engineers, often from several countries.

The revisions and additional material in this edition bring the story up to date. They include an extra chapter on recent particle physics, with a section relating these discoveries to cosmology. More attention has been given to nuclear reactions, the exploration of which at higher energies with enhanced detectors also promises to provide links to particle physics and cosmology. The more mathematical material has been transferred or confined to Appendices, including an improved treatment of the bound state of the deuteron, and a simple introduction to SU(2) and SU(3). In response to several requests, a comprehensive set of references has been prepared which reinforces interest in the physicists involved as well as the concepts and relationships that they study, and provides an historical perspective.

Painswick   E. J. B.
June 1987

Preface to first edition

Nuclear physics is both popular and important. Its popularity is evident in schools and continues right through to postgraduate studies. Its importance is evident not only in the emphasis on fundamental research with very high energy machines, but also in numerous applications. Most of these, ranging from nuclear power stations to the use of nuclear radiation in medicine, need to be seen in relation to the environment. Research into elementary particles continues to produce remarkable results that deepen our understanding of the forces of nature and show how limited are our 'common-sense' conceptions of space and time. It is not surprising that self-respecting scientists and engineers recognize that they require a basic understanding of nuclear physics.

This volume is intended to be equally suitable for the university entrant specializing in physics and for the chemist, geologist, engineer, or biologist who is taking a course covering basic science. The mathematics required is not very demanding, and the supporting physics is not too difficult for non-specialists, except for a few sections that could be omitted.

The introductory chapters are largely historical, partly in order to present the ideas required to follow the discovery of the nucleus and the rapid development of the subject, and partly to allay the prevalent tendency to divorce physics as a specialized discipline (at times almost an ideology) from physicists, other scientists, and society in general. Later chapters are increasingly arranged in conceptual or logical order, and become more demanding. Many new concepts need to be introduced in order to bring the reader near enough to the frontier of the subject to appreciate some of the excitement and interest surrounding contemporary discoveries. Inevitably, some topics are treated very briefly, and simple derivations are used in the hope of illuminating the subject without unduly misleading the reader. The problems at the end of each chapter include many that develop points from the text, and are of a fair range of difficulty.

The author remembers with gratitude the late Professor G. K. T. Conn, who made many helpful comments, especially on the first three chapters. The other directly related text in this Series, *The properties of nuclei* by Dr G. A. Jones (OPS 12), assumes familiarity with many of the topics treated here, including accelerators and detectors, and Dr Jones has kindly read this text and made valuable suggestions. The author is grateful to the several friends and colleagues who have taken an interest in parts of the text and contributed to the minimization of errors and the improvement of clarity and style. Finally, to Dr D. J. E. Ingram is due much appreciation for his editing of the script and his encouraging remarks, and to Mrs J. Ashdown many thanks for her careful typing.

Acknowledgements

The author wishes to thank the American Physical Society and the authors concerned for permission to reproduce or adapt material for Figs. 3.5, 3.6, 5.6, and 8.3.

E. J. B.

Contents

1. Atoms, electrons, and photons

Introduction

Solids and liquids are readily recognized by the senses, but gases and vapours cannot be appreciated so directly. Leaves rustle and trees bend, and so we talk of the wind. By similar processes of thought we talk of magnetic fields when the compass needle is returned to a stable direction. The attraction of small objects by substances such as glass 'electrified' by friction introduces the idea of electric fields. As our experience widens, our powers of conceptualization increase, and so does our ability to manipulate these concepts by appropriate mathematical techniques, notably the differential and integral calculus and, more recently, numerical methods using computers.

From time to time crucial observations are made or striking syntheses of ideas are propounded, and in retrospect they become landmarks of experience. We may mention only a few: Newton's understanding of gravitation (1687) linking the terrestrial to the celestial; Maxwell's electromagnetic theory (1864) linking electricity, magnetism and light; Einstein's theory of relativity (1905) linking energy and mass and identifying the importance of frames of reference. It would be tempting for some to include Dalton's atomic theory (1803), but that would be an injustice to the very long history of development of the concepts of atoms and molecules. The naming of the electron as the smallest electric charge in electrolysis was the suggestion of Johnstone Stoney (1894) but the identification of the definitive properties of the corresponding particle is generally attributed to J. J. Thomson (1897). The introduction of the quantum or photon, on the other hand, is without question the work of Planck (1900), but it was not much appreciated until Einstein (1905a) used the photon to explain the photoelectric effect. Planck's work has as its counterpart the work of de Broglie (1924) linking the wave and particle aspects of matter, which initiated the development of quantum mechanics.

The identification of the nucleus of the atom by Rutherford (1911) is one of the great landmarks in science. It was followed by the recognition of the proton and neutron as constituents of nuclei and the identification of the strong nuclear interaction that binds together these particles in the nucleus, and also accounts for the scattering of neutrons by protons. The type of radioactivity characterised by the emission of beta particles, i.e. high-energy electrons (and neutrinos), could not be explained in terms of the strong nuclear interaction and led to the identification of the weak nuclear interaction. There were then four distinct recognized interactions whereby all known forces were believed to be able to be explained; (i) strong nuclear, (ii) electromagnetic, (iii) weak nuclear, and (iv) gravitational interaction.

The most recent and impressive landmark in science has been the unification of two of these four interactions. The electromagnetic interaction and the weak interaction have been shown to have an underlying unity, at first postulated in terms of a theory based on symmetry (1960), later encouraged by supporting experimental evidence (1973), and recognized in 1979 with the award of the Nobel Prize to the originators, Glashow, Salam, and Weinberg. The crowning achievement of this chapter in particle physics was the experimental identification in 1983 of the W and Z particles, with properties predicted by this electroweak theory. The 'strong nuclear' force is now recognized to be the nuclear equivalent of the van der Waals force, and arises from the more basic '(strong) interquark force'. Attempts to unify all four basic interactions continue, with relevance in both high energy physics and cosmology.

Side by side with these great conceptual advances there have been both unexpected discoveries and deliberate applications. The study of the nucleus led directly to the development of nuclear power for both military and peaceful purposes, and also to the many uses of radioactivity in medicine, industry, and research.

We shall review in Chapter 2 the events that prepared the way for the discovery of the nucleus, and then consider in Chapter 3 the importance of alpha-particle studies in extending our understanding of nuclear properties. The rapid advances in nuclear physics went hand-in-hand with the development of nuclear accelerators and particle and track detectors, which are described in Chapters 4 and 5. Among the wealth of possible nuclear reactions we emphasize fission and fusion (Chapter 6) for their importance in power generation, and in Chapter 7 the fundamental strong nuclear interaction between neutron and proton is treated, together with a brief account of beta emission and nuclear models. In Chapter 8 an introduction is given to cosmic rays and the fascinations and problems of the strange particles. Finally, in Chapter 9, we introduce quarks, the electroweak unification and the third lepton, and relate particle physics to cosmology.

In this chapter we shall consider briefly the history of the concepts of atoms, electrons, and photons. Thereby the reader will be able to link familiar concepts and experimental methods to the new material to be presented. The scene will also be set for the experiments and ideas, outlined in Chapter 2, which preceded the discovery of the nucleus.

Atoms

The concept of the atom can be traced back to the Greeks, Leucippus and Democritus in the sixth and fifth centuries BC. respectively, but it was the work of the eighteenth and nineteenth centuries AD. which led to its definition as 'the smallest particle of matter that retains the chemical properties of an element'. The word 'chemical' is important, and indicates the context of much of the work that led Dalton in 1803 to postulate atoms. He was thereby able to explain the law of conservation of mass in chemical reactions (Lavoisier 1789) and the law of definite proportions (Proust 1799), and he extended the latter to the law of multiple proportions (1804). Dalton's own words (*New system of chemical philosophy*, 1808) are transparently clear. 'If there are two bodies, A and B, which are disposed to combine, the following is the order in which combinations may take place, beginning with the most simple: namely, 1 atom of A + 1 atom of B . . . , 1 atom of A + 2 atoms of B . . . , 2 atoms of A + 1 atom of B, etc.' We might wish to emphasize 'may take place' since not all that many multiples are in practice 'disposed to combine'.

Significantly, Dalton wrote in 1803 'An inquiry into the relative weights of the ultimate particles of bodies is a subject, so far as I know, entirely new. I have lately been prosecuting this enquiry with remarkable success'. Although it was necessary for the accuracy of these measurements to be improved, and the distinction between atomic weights, molecular weights, and combining weights or equivalent weights had yet to be made, there is no doubt that Dalton's work was a major advance. The suggestion of Prout in 1815 that all elements are made up of atoms of hydrogen as a primordial substance was soon abandoned, but was revived decades later in a new form.

The ordering of elements by their atomic weights about 1860 led to the discovery of periodicities in their physical and chemical properties. In 1871 Mendeléev was reasonably successful in systematizing the known elements in his periodic table, mainly because he was bold enough both to leave gaps for undiscovered elements, and also to make small reversals of order in the belief, in part mistaken, that some weights were inaccurate.

From relative weights of atoms we turn to their sizes. A number of estimates based on the kinetic theory of the viscosity of a gas and the density of the corresponding liquid were reviewed by Kelvin in 1870, and led to values of about 10^{-10} m for the diameter of an atom. As early as 1865 Loschmidt used such estimates of size to deduce the Avogadro constant, i.e. the number of molecules in the gram molecular weight, now known to be 6.023×10^{23} mol^{-1}. From this number we can deduce the masses of individual atoms, ranging from 1.67×10^{-27} kg for hydrogen to 395×10^{-27} kg for uranium.

With sizes and masses confidently measured the validity of the atomic theory of matter was beyond serious challenge, although a number of influential scientists (e.g. Ostwald and Mach) opposed the theory even into the twentieth century. The concept of the atom as indivisible, however, was soon to need revision, as a result of experiments on electrical discharges in gases.

Electrons

Electrostatic attraction phenomena using amber and fur were reported by the Greeks about 600 BC. William Gilbert in the sixteenth century studied such phenomena and coined the word 'electricity' from elektron, the Greek word for amber. In 1733 two kinds of electrification were noted by DuFay, and in 1747 Benjamin Franklin independently introduced the terms 'positive' and 'negative' electricity.

The use of gases at reduced pressures for electrical investigations dates back to 1705, when Hawksbee noticed that light was emitted in the frictional electrification of amber in a partially evacuated vessel. By about 1870 the study of low-pressure discharges between metal electrodes with controlled potential differences was beginning to produce significant results. The associated glowing of the walls of the glass discharge tubes was traced to the impact of 'rays' from the cathode, i.e. the negative electrode. William Crookes and other

British workers suggested that the rays consisted of high-velocity negatively charged particles, but Hertz, Lenard, and others favoured an interpretation in terms of short-wavelength electromagnetic waves. Perrin in 1895 collected the particles in a cylinder connected to an electrometer, and demonstrated their negative charge, thereby supporting the particle hypothesis.

In the process of identifying the characteristics of these 'cathode rays', the successes of Thomson (1894–7) were directly due to the development of better vacuum systems. He was able to deflect the particles with both electric and magnetic fields, whereas electric fields used by earlier workers had had no detectable effect, since the cathode rays produced many ions in the residual gas of the relatively poor vacua and these acted as an electric shield. In one of a comprehensive series of experiments, Thomson (1894) measured the velocity v of the particles of charge e by crossed electric (X) and magnetic (B) fields (Fig. 1.1) acting over the same length of the particle beam and adjusted to give zero deflection; then $eX = evB$, and hence $v = X/B$. The velocity produced by his accelerating voltage was about $3 \times 10^7 \text{ m s}^{-1}$, i.e. about one-tenth of the velocity of light, by far the

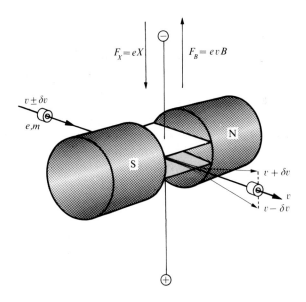

Fig. 1.1. Crossed electric and magnetic fields acting as a velocity selector (Wien filter) for negatively charged particles.

highest velocity ever measured by man up to that time, other than the velocity of light itself.

Thomson used two independent methods to measure the ratio of the mass m to the electric charge e of the particles. In the first he measured the total charge $Ne = Q$ carried by N particles and also their total energy $\frac{1}{2}Nmv^2 = W$. The velocity was then obtained by deflecting the particles in the magnetic field B. The beam follows a circular path of radius R and the force mv^2/R required to constrain a particle to move in this circle is provided by the force evB, produced by the magnetic field,

i.e.
$$mv^2/R = evB,$$

hence
$$v = RBe/m,$$

and
$$W = \frac{1}{2}NR^2B^2e^2/m,$$

Then
$$m/e = R^2B^2Q/2W.$$

In the second method he used the technique of crossed electric and magnetic fields to determine v, and then measured the angular deflection θ (Fig. 1.2) caused solely by the electric field X acting over the length l of the beam. The acceleration perpendicular to v due to X is given by eX/m and acts for a time l/v, producing a transverse velocity at exit from the field given by $v_{tr} = eXl/mv$. Since θ is small it is equal to the ratio of this velocity to the initial velocity of the beam, hence $\theta = eXl/mv^2$. Inserting $v = X/B$ yields $m/e = lB^2/X\theta$.

The values obtained for m/e were essentially independent of the nature of the residual gas in the tube (air, hydrogen, carbon dioxide) and of the material of the electrodes (aluminium, platinum) and were about one-thousandth of the value for the hydrogen ion as found from electrolysis experiments. Similar results were obtained about the

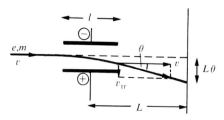

Fig. 1.2. Deflection of a negatively charged particle beam of known velocity by an electric field.

same time by Kaufmann and Wiechert. At first, Thomson (1897) wrote 'the smallness of m/e is, I think, due to the largeness of e [compared with that carried by an ion in an electrolyte] as well as the smallness of m.'

It remained for Thomson to measure the charge on the individual particles, but the cathode rays did not lend themselves to the methods available. He therefore first showed the identity of cathode rays and photoelectrons by measuring m/e for the latter. Then he used a technique developed by C. T. R. Wilson (1900) (later to be of great value in the nuclear track detector known as the cloud chamber), in which ions in a gas act as centres for condensation of supersaturated water vapour. The radius of the drops formed can be estimated from the downward velocity of the cloud. The total weight of the cloud and the charge it carries then give the charge per droplet. The value was evidently equal in magnitude to the (positive) charge carried by the hydrogen ion in electrolysis. The work of Millikan (1909) based on his well-known oil-drop method improved considerably the accuracy of the measurement of the electronic charge. Measurements by Bäcklin (1928) based on X-ray diffraction data and a knowledge of the Avogadro constant N_A were even more precise, and they served to reveal a systematic error in Millikan's value arising from the use of an incorrect value of the viscosity of air. The accepted value of e is -1.60×10^{-19} C. The mass of the electron was estimated by Thomson to be about 1.4×10^{-3} of that of the hydrogen ion m_p. The accepted value of m_e is 9.1×10^{-31} kg, i.e. about $1/1836 \times m_p$, and together with the value of e can be regarded as the characteristic parameters of the electron.

Early theories of atomic structure

Since neutral atoms in a gas under the action of ultraviolet light, or X-rays, produced electrons and positive ions, it was evident that the atom can be divided, in spite of the usual interpretation of its name. The structure of the atom became a matter for speculation, and Thomson in 1899 suggested that there are 'a large number of corpuscles [i.e. electrons] ... spread' in a positively charged space. By 1904 he had developed the idea in terms of '[electrons] in a series ... of concentric shells ... moving about in a sphere of uniform positive electrification', and he explicitly related this to the periodic classification of the elements, an idea first put forward in his 1897 paper on the electron. Measurements of the scattering and absorption of X-rays by gases were interpreted by Thomson, and indicated that the number of electrons per atom was of the order of the atomic weight—in fact it is equal to the atomic number, Z—and therefore most of the mass was in the positively charged part of the atom.

The next significant development in theories of atomic structure was the result of Rutherford's interpretation of alpha-particle scattering in terms of the nucleus (1911), but it is worth noting, in the same volume of *The Philosophical Magazine* that contained Thomson's 1904 paper, the neglected publication of the Japanese physicist Nagaoka (1904) to which Rutherford himself drew attention. Nagaoka compared the atom to the planet Saturn and continued '[the atom] will evidently be approximately realized if we replace these satellites [i.e. Saturn's rings] by negative electrons and the attracting centre by a positively charged particle.' He attempted to explain characteristic atomic spectra, and even radioactivity, but states clearly 'the objection to such a system of electrons is that the system must ultimately come to rest, in consequence of the exhaustion of energy by radiation, if the loss be not properly compensated.'

This objection, arising from the classical theory of a charged particle accelerated by the force constraining the electron to move in a circular orbit (see pp. 59, 74), was avoided by Bohr's assumptions in his quantum model of the atom (Bohr 1913). He postulated non-radiating, 'stationary' orbits, characterized by allowing only those angular momentum values that are whole number multiples of the angular momentum of the smallest orbit. This leads to quantized energies, and hence to transitions from one orbit to another which are accompanied by the emission or absorption of light with characteristic frequencies (or wavelengths). The resulting spectrum agrees remarkably well with the observed line spectrum for atomic hydrogen.

Photons

The suggestion that light consists of particles which enter the eye from luminous or illuminated bodies was current in the circle of Pythagoras about 500 BC. Newton revived the possibility in the seventeenth century but his writings allow that undulatory or oscillatory features are likely to be present. His contemporaries,

Hooke and Huygens, favoured a wave theory, and of course the well-known Newton's rings experiment finds a ready explanation in terms of periodic phenomena or concepts. It is particularly interesting to read Newton's own account of this in his *Opticks* (1704). Early in the nineteenth century, Young and Fresnel gave explanations in terms of wave theory for the three optical phenomena that are commonly quoted as evidence for wave properties, (1) diffraction effects at the boundaries of shadows, (2) interference as, for example, between light from nearby slits illuminated by a common source, and (3) polarization effects. Visible light was shown to have a wavelength of about 5×10^{-7} m, i.e. about 1/2000th of a millimetre, and together with the velocity of light (3×10^8 m s^{-1}) this yields a frequency of about 10^{15} Hz (hertz = cycles per second). Colours were explained by different wavelengths, longer for red ($\simeq 7 \times 10^{-7}$ m) and shorter for blue ($\simeq 4 \times 10^{-7}$ m).

Faraday established a relation between magnetism and electricity, and suspected that there was a relation between electromagnetism and light. He succeeded in 1845 in producing a small but detectable rotation of the plane of polarization of a beam of light passing through lead glass by applying a magnetic field parallel to the direction of the beam. This is now known as the Faraday effect. However, it was not until 1865 when Maxwell developed his four electromagnetic equations, including the new concept of displacement current, that the possibility arose of interpreting light as an electromagnetic wave. Maxwell's equations showed that a plane-polarized wave *in vacuo* consists of oscillating electric and magnetic fields having directions perpendicular both to each other and to the direction of propagation of the wave. For monochromatic light these fields vary in magnitude at any one point in space as $\sin \omega t$, where ω is the angular frequency ($\simeq 10^{15}$ Hz for visible light), and at any one time the fields vary as $\sin(2\pi x/\lambda)$ along the beam direction x, where λ is the wavelength. Their velocity is given by $c = (\varepsilon_0 \mu_0)^{-\frac{1}{2}}$, where μ_0 is the magnetic permeability of a vacuum ($= 4\pi \times 10^{-7}$ H m^{-1} exactly by definition) and ε_0 is the electric permittivity of a vacuum ($= 8.854 \times 10^{-12}$ F m^{-1}). Electric and magnetic measurements confirmed the accuracy of these quantitative predictions, and the experiments of Hertz (1887) with oscillatory electric discharges clearly demonstrated the existence of electromagnetic waves with wavelengths of the order of a metre, i.e. short radiowaves. The elegance, accuracy, and apparent completeness of Maxwell's theory seemed at first to leave little opportunity for developments in the fundamental understanding of the propagation of light and other electromagnetic energy.

There were, however, clouds on the horizon. Maxwell's equations, unlike the laws of dynamics, change their form when 'commonsense' (Galilean) transformations are used to convert from one frame of reference of space and time (x, y, z, t) to another that is moving with a constant relative velocity. For a relative velocity v along common x-axes the Galilean transformation is simply $x' = x - vt, y' = y, z' = z$, and $t' = t$. But careful experiments had shown that electromagnetic processes are independent of the inertial frame of reference in which the observations are made. This was one of the problems that required the theory of relativity presaged by Lorentz and propounded by Einstein (1905b). The Lorentz transformation for such a relative velocity yields $x' = \gamma(x - vt)$, $y' = y$, $z' = z$, and $t' = \gamma(t - vx/c^2)$ where $\gamma = (1 - v^2/c^2)^{-\frac{1}{2}}$. Under this transformation the Maxwell equations, and indeed all accepted 'laws' of physics, are covariant, i.e. they retain the same form or appearance in the two systems.

Furthermore, there was the unsolved problem of the wavelength distribution of the radiation emitted by a hot 'black' body, as measured, for example, by Lummer and Pringsheim (1897–1900). Rayleigh (1900) and Jeans (1909) applied classical concepts and Maxwell–Boltzmann statistics to the emitting oscillators of the body and obtained agreement for the longest wavelengths, but predicted infinite emission at the shortest wavelengths.

In 1900 Planck broke away from the continuity of thought of classical physics in his explanation of black-body radiation. He first obtained an accurate empirical formula and then recognized the necessary assumption, namely, that the oscillators of frequency v could emit energy only in 'energy elements', now known as quanta, of magnitudes $hv, 2hv, \ldots, nhv$, where n is a whole number and h is a constant, the Planck constant, 6.626×10^{-34} J s. Even for light with $v \simeq 10^{15}$ Hz, hv is a minute amount of energy—for green light it is about 2.5 eV, i.e. the kinetic energy of an electron accelerated from rest across a potential difference of 2.5 V. When it radiated, the oscillator was considered to radiate all of its energy nhv, but each oscillator was assumed to absorb energy in a continuous manner, and the radiation field around the body was not assumed to be populated with quanta. The observed form of the variation with wavelength of intensity per unit wavelength interval was accurately fitted by Planck's equation simply by finding the appropriate magnitude of h

(Fig. 1.3). The energy density per unit wavelength interval, at a given wavelength λ, is given by

$$E_\lambda = \frac{8\pi}{\lambda^4} \frac{h\nu}{\exp(h\nu/kT) - 1}$$

where k is the Boltzmann constant and T the absolute temperature. Notice that for values of h approaching zero, Planck's formula reduces to the classical values of Rayleigh and Jeans, $E_\lambda = 8\pi kT/\lambda^4$.

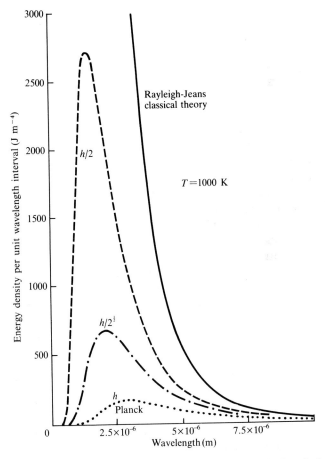

Fig. 1.3. Relation between classical and quantum theories of black-body radiation. As $h \to h/2^{1/2} \to h/2 \to 0$, Planck's expression transforms to that of Rayleigh and Jeans.

The photoelectric effect (a by-product of the experiments of Hertz 1887 and first studied by Hallwachs 1888) is the emission of electrons from matter when it is illuminated by light or other electromagnetic radiation of adequately short wavelength. Einstein (1905a) interpreted the threshold wavelength λ_{th} as that of a photon which has just sufficient energy $h\nu_{th} = hc/\lambda_{th}$ to remove from the material the electron with the smallest binding energy $e\varphi$, usually referred to as the work function. For ultraviolet light $h\nu$ is about 5 to 10 eV, and any photon of energy greater than $h\nu_{th}$ ($=e\varphi$) can yield photoelectrons with kinetic energies up to a maximum value $T_e(\max) = h\nu - e\varphi$. The success of this equation quickly established the need to consider electromagnetic radiation in terms of photons whenever the energy involved in the process is of the order of $h\nu$.

For many people, it is a great temptation to try to form a mental picture of the photon. Commonly, the photon is conceived as being extremely small in size. It then comes as a surprise to learn that interference phenomena can be recorded and built up, photon by photon, for two apertures several metres apart as in the Michelson stellar interferometer. The question arises, does each photon sample both apertures? Another common suggestion is that light is emitted and absorbed in quanta yet propagated as waves, but in order to investigate its propagation it is necessary to interact with the light and it interacts ultimately in terms of quanta. Wave phenomena require for their interpretation a knowledge of the phase of the wave, and the accuracy with which this phase can be expressed or measured increases with the number of photons involved. The phase is essentially indeterminate for one photon.

It must be recognized, however, that the lack of sophistication and caution in the mental pictures entertained by many pioneering physicists did not prevent, and might even have facilitated, their making very important experimental and theoretical advances.

Problems

A table of physical constants and conversion factors is provided at the end of this book (p. 203).

1.1. The velocity of electrons in one of J. J. Thomson's experiments in 1897 was found to be 3×10^7 m s^{-1}. What can be deduced about the accelerating potential difference that he used?

1.2. There are five interrelated parameters in the design of Thomson's apparatus depicted in Figs. 1.1 and 1.2, namely, θ, v, l, X, and B. Suggest suitable values for these quantities. Comment on the expected accuracy in θ, bearing in mind the size of the electron beam. Hence consider the accuracies of v, l, X, and B, and deduce the expected accuracy in e/m.

1.3. In a television tube, a beam of electrons of energy 1 keV is deflected through an angle of 45°. Calculate and comment on the strength and linear extent, l, of (a) an electric field, X, and (b) a magnetic field, B, able to produce such a deflection.

1.4. A candle flame can be seen at night at a distance of 2 km by the unaided eye. If it emits visible light at the rate of 0.02 J s^{-1}, what is the approximate number of photons entering the eye per second? State clearly your assumptions and approximations.

1.5. What is the wavelength of a beam of ultraviolet light if its photon energy is 10 eV? If the beam produces photoelectrons from iron, which has a work function of 4.72 eV, what is the maximum energy and the maximum velocity of the electrons?

1.6. Calculate the percentage increase in the energy density per unit wavelength interval E, as given by the Planck radiation formula, which would be produced by a 1 per cent decrease in h at $\lambda = 3 \times 10^{-6}$ m for $T = 1000$ K.

2. Nuclear physics before the nucleus

Radioactivity

Nuclear physics was born in 1896, ten years before Rutherford fathered the nucleus. The occasion was the accidental discovery of radioactivity by Becquerel, during experiments on the X-rays found by Röntgen in the previous year. In this chapter we consider how these X-ray experiments led to the identification of alpha, beta, and gamma rays, and how the laws of radioactive decay were formulated and the radioactive series explored. The properties of alpha and beta particles are then considered, together with the characteristics of their emission. Finally, we introduce positive ions and isotopes, and the important relation between mass and energy. This chapter, therefore, sets the scene for the discovery of the nucleus, presented in Chapter 3.

In 1895 Röntgen noticed that electrons in cathode-ray tubes made the glass walls luminesce, i.e. produce visible light, but some compounds such as zinc sulphide and barium platinocyanide gave a much more intense effect than glass. He also discovered that a discharge tube totally enclosed in black cardboard still produced a strong luminescence on a nearby piece of paper coated with barium platinocyanide. Röntgen identified the penetrating rays as X-rays, 'for the sake of brevity', and noted that they also fogged photographic plates. Their nature as electromagnetic waves of very short wavelengths (about 10^{-10} m) was indicated by Haga and Wind (1899) using a tapered slit to produce a diffraction pattern, and established by Friedrich, Knipping, and von Laue (1912) using a crystal as diffraction grating—thereby inaugurating X-ray crystallography with its manifold applications.

Henri Becquerel (1896) attempted to find a relation between X-ray luminescence and the 'fluorescence' in potassium uranyl sulphate, i.e. the production of light of a particular wavelength by the salt during exposure to light of a shorter wavelength, e.g. sunlight. A photographic plate was wrapped in black paper, and a thin crystal of uranium salt placed on it. Exposure to sunlight and subsequent development revealed local darkening, as if X-rays had been produced by the sun's action! Three sunless days in February 1896 proved to be a blessing in disguise, for it was found that prepared plates and crystals stowed unexposed in a drawer showed remarkable blackening when developed. It was evident that rays, rather like X-rays in their action, were emitted by the uranium crystals, without the stimulus of the sun's rays and, in Becquerel's words, 'the time of persistence is infinitely greater than that of the visible radiations emitted by such bodies.'

The investigation of 'radioactivity'—the word was coined by Marie Curie in 1898—was pursued in several countries. Some of the 'rays' were readily stopped by 0.02 mm of aluminium, and Rutherford (1899) called these alpha-rays. They were later shown to be doubly ionized helium atoms. A second type required a stopping thickness of about 1 mm of aluminium and were termed 'beta-rays' by Rutherford (1899), soon to be identified by Becquerel (1899) as fast electrons because of their negative charge and e/m value.

A third type of radiation was recognized by Villard (1900) and came to be known as gamma-rays in 1903. These rays were not deflected by a magnetic field and had considerable penetrating power. Satisfactory evidence of their electromagnetic nature was not obtained until 1914 when Rutherford and Andrade (1914) showed that gamma rays could be diffracted by crystals, and had wavelengths shorter than X-rays. Thus the wavelength of 10 keV X-rays is 1.24×10^{-10} m (1.24 Å), and for 1 MeV γ-rays it is 1.24×10^{-12} m.

After the three sunless days of February 1896, Becquerel concentrated his attention on the new radiation from uranium and soon showed that it was independent both of the state of chemical combination of the element and of its temperature. Once it was realized that the rays readily ionize gases, since they could discharge an electroscope, new tools for their study became available. Marie Curie and her husband Pierre Curie entered the scene in 1898, and quickly showed that thorium had similar radioactive properties. From pitchblende they separated a new substance 'whose activity (i.e. disintegrations per second) is about 400 times as great as that of uranium'—'we propose to call it polonium' (after Marie's homeland, Poland). An even more active element, radium, was later identified by the Curies, and confirmed by enormously laborious chemical separations that yielded 0.1 g of radium chloride from one ton of pitchblende residues.

The time-dependence of radioactivity was conjectured by Becquerel (1901) after Crookes (1900) had shown that the active substance in uranium nitrate could be chemically separated from the uranium. Becquerel argued that since all natural uranium salts had a characteristic 'equilibrium' activity proportional to the uranium present, the chemically 'deactivated' uranium would recover its equilibrium activity—and it did! But Becquerel had to wait eighteen months for what he described as 'complete recovery'.

Rutherford and Soddy (1902), working with thorium under rather special conditions (see Problem 2.4), obtained results similar to those shown in Fig. 2.1 (on elevation to the peerage, Rutherford had a version of this diagram incorporated in his escutcheon). Thorium X was the active component chemically separated from the natural thorium, and its decay curve had as its complement the curve of recovery of the deactivated thorium. It was shown that the reactivated thorium can once again have its thorium X activity separated and will again recover. The explanations soon followed:

1. Emission of alpha-particles by thorium is accompanied by a change of element, i.e. there is a transmutation of the chemical element—the dream of the alchemists come true.

2. The element formed, thorium X, is itself strongly radioactive, and can be separated chemically from thorium.

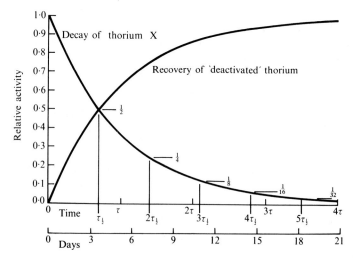

Fig. 2.1. Decay of the chemically-separated active component of thorium, and recovery of the 'deactivated' thorium. (For the 'mean life', $\tau = 1/\lambda$, see problem 2.3.)

3. Thorium X decays to a very weakly radioactive product.

4. Natural thorium robbed of thorium X decays to produce more thorium X until equilibrium is attained, i.e. until the rate of production of thorium X equals its inherent decay rate.

The variation of the activity A of thorium X with time is a simple exponential,

$$A = A_0 \exp(-\lambda t),$$

where λ is the radioactive constant of thorium X and A_0 is the activity at time $t = 0$. This may be derived from the basic assumption that λ is the probability per unit time that a given nucleus of thorium X will decay. For N nuclei at time t,

$$dN/dt = -\lambda N.$$

Integration of $dN/N = -\lambda dt$ yields $\ln N = -\lambda t + \text{constant}$. At $t = 0$, $N = N_0$ and the constant is $\ln N_0$ hence $N = N_0 \exp(-\lambda t)$. But the activity A, i.e. the number of disintegrations per second, is λN, and $\lambda N = \lambda N_0 \exp(-\lambda t)$ or $A = A_0 \exp(-\lambda t)$, as required.

The accepted unit of radioactivity, adopted internationally, is the becquerel,

$$1 \text{ becquerel} = 1 \text{ Bq} = 1 \text{ disintegration per second.}$$

This is a very small quantity and many people still use the curie, introduced in 1910 as the activity of the radon found in radioactive equilibrium with one gram of radium. Radon is a gaseous radioactive element, $_{86}\text{Rn}$, sometimes called emanation, $_{86}\text{Em}$. The curie was defined officially in 1948 as exactly

$$1 \text{ curie} = 1 \text{ Ci} = 3.7 \times 10^{10} \text{ disintegrations per second.}$$

This is quite a large quantity!—sources for laboratory demonstrations are often a few μCi. Note that $1 \text{ MBq} \simeq 27 \mu\text{Ci}$.

For thorium X, the radioactive constant determining the rate of decay is $\lambda = 2.20 \times 10^{-6} \text{ s}^{-1}$. A more common characteristic constant used to express radioactive decay rate is the half-life, $\tau_{\frac{1}{2}}$, which is simply the time for N_0 nuclei of the species to decay to $\frac{1}{2}N_0$. Since $N_0/2 = N_0 \exp(-\lambda \tau_{\frac{1}{2}})$,

$$\tau_{\frac{1}{2}} = (\ln 2)/\lambda = 0.693/\lambda.$$

For thorium X, $\tau_{\frac{1}{2}} = 3.64$ days (Fig. 2.1).

The form of the recovery of activity by the deactivated thorium is $A_2 = A_1(0)\{1 - \exp(-\lambda t)\}$ and follows readily from the assumptions:

Thorium $\rightarrow$	Thorium X $\rightarrow$	Product
$\lambda = \lambda_1$	$\lambda = \lambda_2$	$\lambda \rightarrow 0$
$N_0 = N_1(0)$	$N_0 = 0$	$\tau_{\frac{1}{2}} \rightarrow \infty$
$N_t = N_1$	$N_t = N_2$	

At any time t, the rate of production of thorium $X = \lambda_1 N_1$ and the rate of decay of thorium $X = \lambda_2 N_2 = A_2$. Hence

$$\mathrm{d}N_2/\mathrm{d}t = \lambda_1 N_1 - \lambda_2 N_2.$$

Since thorium decays very slowly (half-life $= 1.39 \times 10^{10}$ years) its activity $A_1 = \lambda_1 N_1(0) \exp(-\lambda_1 t)$ is essentially constant during the period of the experiment. Then $\mathrm{d}N_2/(A_1 - \lambda_2 N_2) = \mathrm{d}t$, or $\ln(A_1 - \lambda_2 N_2) = -\lambda_2 t + \text{constant}$; at $t = 0$, $N_2 = 0$ and therefore the constant $\ln A_1(0) \simeq \ln A_1$. It follows that

$$A_2 = \lambda_2 N_2 = A_1\{1 - \exp(-\lambda_2 t)\}.$$

The commonsense interpretation of this result is evident since the fractions, had they not been separated, would together have had a constant activity $A_1 (= A_1(0))$, i.e.

$$A_1 = A_1 \exp(-\lambda_2 t) + A_1\{1 - \exp(-\lambda_2 t)\}$$

natural thorium = thorium X + deactivated thorium

It often happens in physics that simplifying assumptions lead to important advances in understanding. Rutherford and Soddy realized that more than three components were involved in the thorium decay scheme, but the simple model of thorium, thorium X, and final product works well. The detailed picture revealed by later studies appears at first so very different that it provides a useful problem (see problem 2.4) for the reader to explain why the early experiments produced the results they did, and what reformulation is required of the simple assumptions of only three components. In Fig. 2.2 the full sequence of radioactive changes is given for the series that contains thorium. Three radioactive series are found in nature, characterized by their atomic-mass numbers (1) $A = 4n$ (thorium series), (2) $A = 4n + 2$ (uranium series), and (3) $A = 4n + 3$ (actinium series). Each ends with a stable isotope of lead. The fourth series with $A = 4n + 1$ is not

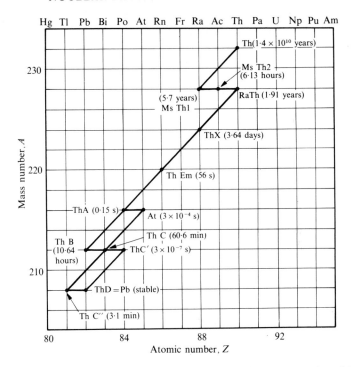

Fig. 2.2. The thorium radioactive series. Diagonal transitions correspond to alpha-emission ($\delta A = -4$, $\delta Z = -2$) and horizontal transitions correspond to (negative) beta-emission ($\delta A = 0$, $\delta Z = +1$).

found in nature but has been produced artificially—the neptunium series (see p. 108).

Types of radiation—alpha, beta, gamma

The identification of alpha-rays was not such an easy matter as the pin-pointing of beta-rays by their e/m value. Deflections of alpha-particles by magnetic and electric fields yielded a ratio of positive charge to mass which was about one-half of that of the hydrogen ion. Rutherford and Geiger (1908a, b) found the charge to be $+2e$, strongly suggesting that the alpha-particle is a doubly ionized helium atom of mass four times that of the hydrogen ion. Their experiment to determine the charge on the alpha-particle was the first in which the Geiger counter (see p. 80) was used, and deserves more attention.

An alpha source in an evacuated enclosure (Fig. 2.3(a)) subtended a known solid angle Ω_1 at a small aperture A covered by a thin mica sheet, which allowed the particles to enter a cylindrical volume V at a pressure of a few torr, 1 torr = 1 mmHg equivalent. A thin axial wire was maintained at a potential of about 1300 V with respect to the cylinder, and the entry of a single particle produced a noticeable flick of the needle of the attached quadrant electrometer. The number n emitted into the small solid angle Ω_1 in a given time was therefore counted directly. For a much larger solid angle Ω_2 the charge Q in the same given time was measured directly in a simple apparatus (Fig. 2.3(b)). A strong magnetic field perpendicular to the axis of the alpha-beam prevented beta-particles from the source reaching the shallow cup collector C, and prevented the escape from C of secondary electrons produced by the alpha bombardment. The aluminium foil stopped recoil atoms from the source. The charge q on each particle was then given by $Q\Omega_1/n\Omega_2$, and was unambiguously $+2e$.

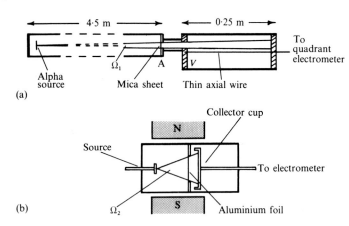

Fig. 2.3. Determination of the electric charge of the alpha-particle by Rutherford and Geiger. (a) Direct count, by Geiger counter, of rate for small solid angle. (b) Current flowing for large solid angle.

An elegant experiment in confirmation of the identification of the alpha-particle as a He^{2+} ion was performed by Sir James Dewar (1908) and repeated by Rutherford and Royds in 1909. Alpha-rays from a source sealed by a thin window entered an evacuated glass vessel terminating in a capillary tube containing two electrodes. After a number of days the gas produced was compressed into the capillary and an electric discharge revealed the characteristic spectrum of helium gas.

The ionization produced by alpha-particles during their passage through a gas was investigated by W. H. Bragg (1904) using the arrangement shown schematically in Fig. 2.4(a). The ionization between the two grids, A, B, about 1 mm apart, was measured by the electric current flowing between A and B when a sufficiently large voltage was applied to collect all the ions initially produced in that region by the beam of alpha-particles. An example of the values observed is shown in Fig. 2.4(b). As the alpha-particles ionize the gas and lose energy the probability of ionization increases—later work showed that the energy loss per unit path length is approximately inversely proportional to the particle energy.

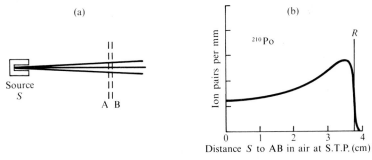

Fig. 2.4. Ionization produced by alpha-particles in air. (a) Thin ionization chamber AB. (b) Variation with distance from source, showing range R.

Particles from a given source usually have a well-defined range R; some sources have several groups each with a characteristic range (see pp. 43 ff). The small variation in the range of particles arises from the 'straggling' associated with the statistical fluctuations in the loss of energy during the individual ionization processes.

Careful measurement of alpha-particle ranges and velocities led Geiger to the approximate relation T_α(in MeV) $= 2.12\,R^{2/3}$ where T_α is the initial energy of the alpha-particle and R is the range in centimetres of air at standard temperature and pressure (S.T.P.). Rutherford suggested in 1907 that there might be a relation between the range R of an alpha-particle and the radioactive constant λ of the corresponding alpha-emission. In 1911 Geiger and Nuttall proposed the rule, or relation,

$$\ln \lambda = A \ln R + B,$$

where A and B are constants; using Geiger's range relation this can be rewritten

$$\ln \lambda = A_1 \ln T_\alpha + B_1 .$$

The interpretation of this relation by Gurney and Condon, and independently by Gamow, in 1928, was one of the first significant successes of wave mechanics in nuclear physics (see p. 46 ff and Appendix B).

Advances in the study of beta-particles prior to Rutherford's discovery of the nucleus were less evident than those involving alpha-particles. The identification of beta-particles as fast electrons had been quickly established by Becquerel (1899) from their e/m value. For some years their energy spectrum was a matter of dispute—were they monoenergetic as suggested by von Bäyer, Hahn, and Meitner (1911), or with a range of energies, or both? A convincing demonstration of the continuous energy spectrum of the beta-particles emitted by nuclei was provided by Rutherford and Robinson (1913), using a magnetic deflection method similar to that of Thomson (see pp. 5 ff, 24 ff). But there are also monoenergetic groups of electrons, for certain radioactive substances, with intensities that are usually only a few per cent of the total radioactivity (see Fig. 2.5 for recent results). These line spectra of electrons are found in association with not only the continuum of beta-particles but also with the discrete alpha-particle spectra. They are now known to be the result of a direct transfer of (quantized) nuclear energy from an excited nucleus to one of the atomic electrons, leaving the nucleus in a lower excited state or in its

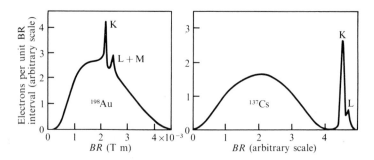

Fig. 2.5. Beta-ray spectra with K, L, and M conversion electrons. BR is the 'magnetic rigidity' of the electrons, i.e. the product of the applied magnetic flux density B in the magnetic beta-spectrometer and the radius R of the circular path of the electron. $BR = p/e$ where p is the momentum of the electron.

ground state. Unfortunately, they are commonly referred to as 'conversion electrons' because of the early suggestion by Rutherford (Rutherford *et al.* 1914) that they are produced by gamma-rays emitted from the excited nucleus which are converted via a photo-electric effect with the atomic electrons. This can happen, but is very improbable.

The continuous range of energies in the beta-particle spectrum presented formidable problems. When a body of well-defined mass m_1 spontaneously breaks up into two particles each with definite mass m_2, m_3, the available kinetic energy T is shared in a unique way. This follows from the conservation laws of linear momentum and energy, thus

$$m_2 v_2 = m_3 v_3 = p \quad \text{and} \quad T = \tfrac{1}{2} m_2 v_2^2 + \tfrac{1}{2} m_3 v_3^2$$

(in the non-relativistic case). Then

$$T = (p^2/2)(1/m_2 + 1/m_3) \quad \text{and} \quad T_2 = T m_3/(m_2 + m_3),$$

with $T_3 = T m_2/(m_2 + m_3)$. The range of energies for the beta-particles therefore appears to rule out two-particle break-up. Could there be two or more betas in each disintegration? This was ruled out by a number of considerations, including chemical identification of the initial and final nuclei which indicated a unit change of nuclear charge. Could there be a continuous range of energy states (and therefore masses) for the initial or final nucleus? That would imply a continuous range of gamma energies, and such a gamma spectrum from nuclei is never observed; all gamma spectra consist of discrete lines. Desperate attempts to solve these problems included a proposal that energy and momentum may not be conserved. In 1931 Fermi developed the suggestion by Pauli that a curious new particle, the neutrino, is emitted with the electron. This particle was assumed to have zero rest-mass, and therefore travels with the velocity of light; it is electrically neutral and extremely difficult to detect (pp. 72 ff). The theory was strikingly successful in its explanation of the beta-particle spectrum, and the neutrino has now been detected by several distinct methods. (See pp. 127, 169 for the current theory.)

Very little was known about gamma-rays in the first fifteen years after the discovery of radioactivity. It was noticed that they were always associated with alpha-emission or beta-emission, i.e. a pure gamma-emitter was not found in nature. Once their energies could be measured in the mid-1920s the opportunity arose to correlate them with alpha-particle spectra (see pp. 43 ff). Their interactions with matter led to new discoveries, e.g. the positron (see pp. 75, 140).

Positive ions

The discovery of the electron in gaseous discharges prompted a search for a corresponding positively charged particle. Goldstein in 1886 observed luminous rays repelled from the anode and passing through holes in the cathode. About the turn of the century, Wien measured the charge-to-mass ratio q/M for these particles by means of electric and magnetic delections and the values were even less than that for hydrogen ions in solutions, suggesting particles thousands of times heavier than electrons. Wien soon showed that, assuming positive charges of the same magnitude as the electron charge, the masses corresponded to the atomic masses of the gas in the discharge. The lightest positive particle corresponded to the hydrogen atom or ion, and as late as 1914 Rutherford still described it as 'the positive electron', a terminology now reserved for the positron. The name proton was not used explicitly for the hydrogen ion until 1920.

J. J. Thomson began to develop apparatus for more precise measurements of q/M for positive rays in 1907 and by 1911 was able to separate two distinct components in neon, one to be assigned a mass of approximately 20 and the other 22, on a scale with oxygen set at 16. His apparatus for e/m of the electron used electric and magnetic fields perpendicular to each other and to the electron beam. As shown in Fig. 2.6, for positive ions he used an electric field X which was parallel to a magnetic field B. It was shown on p. 6 that the angular deflection θ produced by X alone is given by $\theta = qXl/mv^2$,

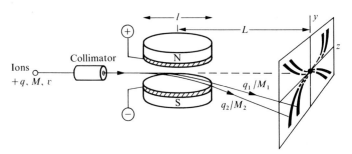

Fig. 2.6. Schematic view of J. J. Thomson's mass spectrograph.

where l is the length of the (undeflected) beam over which X acts and v is the velocity of the particle of mass m. The corresponding displacement y on the screen, distant L from the centre of the field region, is simply $L\theta$ (Fig. 1.2) or, for a particle of mass M, $y = (XlL)(q/M)/v^2$.

The corresponding displacement z on the screen for the magnetic field B acting alone is given for small deflections by $(BlL)(q/M)/v$. This follows readily from Fig. 2.7, since $z \simeq L\varphi$, $\varphi \simeq l/R$ and R is obtained from $Mv^2/R = qvB$. It is clear that, since y is proportional to $1/v^2$ and z is proportional to $1/v$, the curve produced on the screen by both X and B acting together for ions with a range of velocities will be part of the parabola

$$z^2 = (lLB^2/X)(q/M)y,$$

and there will be a separate curve for each value of q/M. Since the maximum velocity for a discharge tube with voltage V is given by $qV = \frac{1}{2}Mv^2$, the minimum value of y is $XlL/2V$.

Aston, working with Thomson, managed by means of gaseous diffusion to effect a partial separation in bulk of the two types of neon. In the same year (1913) Soddy collated the results of many workers and described radioactive species of identical chemical properties as isotopes, i.e. they occupied the same place (Greek: *topos*) in the periodic classification of the elements. Thereby was brought together the important realization that atoms of the same chemical element, containing the same number of electrons, could nevertheless have different masses, whether light atoms as in the case of neon or heavy as found among radioactive elements. Precise measurements of isotopic

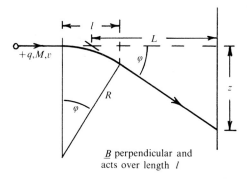

Fig. 2.7. Deflection of charged particle beam by a magnetic field.

masses are of considerable importance in studies of nuclear structure and reactions (see pp. 95 ff).

Mass–energy relation

The history of most advances in science is a tapestry of interacting ideas. Energy was considered to be manifested in various forms in classical physics, the total amount in a closed system being conserved and able to be made any value, i.e. energy was 'continuous'. But in problems requiring quantum physics it is found to appear in discrete amounts, yet still in various forms and still conserved. By the early nineteenth century, mass had come to be regarded as conserved, and in terms of atoms and molecules it was therefore treated as a definite, immutable property. When an atom of mass m_1 was split into an electron m_2 and a positive ion m_3 it seemed natural to assume that the total mass was still conserved, i.e. $m_1 = m_2 + m_3$. But it was later to be discovered that this equation is not exact (see p. 29).

Abraham (1903) and others made many attempts to account for the electron mass entirely in terms of the amount of energy associated with its electromagnetic field but they all failed. If the mass of an electron with a uniform spherical charge distribution is taken to be entirely electromagnetic in origin, the mass will be inversely proportional to its radius, r_0, since the electrostatic or Coulomb energy is given by $E_C = e^2/4\pi\varepsilon_0 r_0$. Lorentz applied to moving electrons the so-called Lorentz–Fitzgerald contraction, first proposed in 1893 in order to account for the Michelson–Morley experiment (1881–7) which had failed to detect movement of the Earth relative to the 'fixed ether' (Michelson and Morley 1887). Note that the length contraction, $l = l_0(1 - v^2/c^2)^{1/2}$, was claimed to follow from Maxwell's electromagnetic equations but this has not been proven satisfactorily. For a moving electron, Lorentz obtained the relation $m = m_0(1 - v^2/c^2)^{-1/2}$ where m is the mass at speed v, m_0 is its rest mass, and c is the speed of light. The argument used for this mass increase is now also generally agreed to be fallacious, but the formula was correct and came at a convenient time to account for the results of Kaufmann (1901) and Bucherer (1908) on the variation with speed of e/m for an electron. Kaufmann used a beam of beta particles in parallel electric and magnetic fields (cf. Thomson's positive-rays apparatus) to produce traces that departed from parabolas for the highest speeds ($\simeq 0.94c$) at which the mass was $\simeq 3.1 m_0$. Bucherer used a velocity filter with crossed electric and magnetic fields, and deflected the resultant monoenergetic beam with the same magnetic field acting alone. The elegant geometry allowed simultaneous measurement of a wide range of speeds, and in his work he used beta particles with speeds up to $0.68c$. Numerous experiments and machines (e.g. nuclear accelerators) have since been based on the validity of $m = m_0(1 - v^2/c^2)^{-1/2}$ for any particle, charged or neutral, and thereby it has been tested to a high order of accuracy.

Whereas Lorentz viewed his derivation of the Fitzgerald contraction in terms of purely formal space–time transformations, Einstein (1905b) was bold enough to make the revolutionary suggestion that the Lorentz transformation expressed relations between *physical* space and time. The absolute frame of reference of the ether was abandoned, and the first postulate of special relativity emerged, namely, that physical observations and their corresponding theoretical interpretations (i.e. physical 'laws') are equally valid for all observers moving with constant linear speeds relative to each other. (General relativity, on the other hand, is concerned with accelerated frames of reference or observation.) The second postulate is that the speed of light *in vacuo* is constant for all observers. The variation of mass with speed follows directly from these postulates, and independently of whether the particle is charged or neutral, and of whether or not its mass is electromagnetic in origin. (This follows from defining momentum as the product of the scalar 'mass' of a particle and its ordinary velocity **v**, and requiring momentum to be conserved in an isolated dynamical system. An alternative approach, aesthetically more simple, is based on the four-dimensional continuum, x, y, z, ict, for which momentum is defined as the product of 'rest mass', an invariant scalar, and the 4-vector velocity, an invariant 4-vector with components, $v_x(1 - v^2/c^2)^{-1/2}$, $v_y(1 - v^2/c^2)^{-1/2}$, $v_z(1 - v^2/c^2)^{-1/2}$ and i$c(1 - v^2/c^2)^{-1/2}$, where $v^2 = v_x^2 + v_y^2 + v_z^2$. The relativistic form for momentum is then explicitly related to our revised view of space and time, and does not depend on what appears to be a new property of matter, namely that its mass increases with speed.)

Another extremely significant derivation by Einstein relates the increase in mass to the increase in kinetic energy of the particle, or more generally $m_0c^2 + T = mc^2 = E$ where T is the kinetic energy and E is the total energy, i.e. the sum of T and the rest-mass equivalent energy m_0c^2.

A simple derivation of $T = (m - m_0)c^2$ can be made starting from $m = m_0(1 - v^2/c^2)^{-1/2}$. The change in kinetic energy dT produced by a force F acting on a body moving through a distance dx is simply dT

$= F\,\mathrm{d}x$. But force is the rate of change of momentum $F = \mathrm{d}(mv)/\mathrm{d}t$ and hence $\mathrm{d}T = \mathrm{d}(mv)\mathrm{d}x/\mathrm{d}t = v\,\mathrm{d}(mv) = v^2\mathrm{d}m + mv\,\mathrm{d}v$. The relation $m = m_0(1 - v^2c^2)^{-1/2}$ is conveniently re-expressed as $m^2(c^2 - v^2) = m_0^2c^2$, and differentiation yields

$$(c^2 - v^2)\mathrm{d}m - mv\,\mathrm{d}v = 0.$$

But

$$v^2\mathrm{d}m + mv\,\mathrm{d}v = \mathrm{d}T,$$

as shown earlier, hence $\mathrm{d}T = \mathrm{d}mc^2$, and for a particle accelerated from rest $T = (m - m_0)c^2$. (The 4-vector approach notes that the 'time' component of the 4-vector momentum is

$$P_4 = im_0c(1 - v^2/c^2)^{-\frac{1}{2}}$$

and is conserved in particle collisions when summed over all the particles. Expansion by the binomial theorem yields

$$P_4 = im_0c(1 + \tfrac{1}{2}v^2/c^2 + \tfrac{3}{8}v^4/c^4 + \ \ldots)$$

and the most important variable term is $(i/c)(\tfrac{1}{2}m_0v^2)$, or $(i/c)T$. It is evident that the total energy E is $P_4(c/i)$ and, for small speeds, $(v \ll c)$, we have $E = m_0c^2 + T$.)

The magnitude of relativistic increases in mass with speed and the corresponding kinetic energies for electrons (in MeV, mega electron volts) and protons (in GeV, giga electron volts) can be seen in Fig. 2.8. It is evident that none of the particle speeds observed in physics before 1900 were large enough to produce detectable changes of mass. Furthermore, only electrons are readily accelerated to high enough

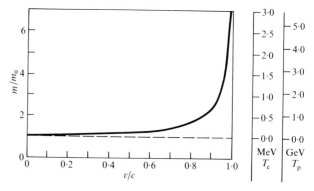

Fig. 2.8. Relativistic increase of mass with speed (with magnitudes for electrons and protons).

speeds for the fractional change in mass to be measured—even so the potential differences required are hundreds of thousands of volts. Fortunately, the energies of beta-particles range up to one or two million electron volts and they enabled Kaufmann and Bucherer to measure the mass variation.

The equivalence of mass and energy is not restricted to kinetic energy. If energy Q needs to be injected into a system (e.g. an atom) to split it into two parts (e.g. an electron and a positive ion) then in the simplest case where the two parts are separated, but at rest, the sum of their rest masses $m_2 + m_3$ will be equal to the rest mass of the atom m_1 plus the mass equivalent Q/c^2 of the injected energy, i.e. $m_1 + Q/c^2 = m_2 + m_3$. The energy Q is referred to as the binding energy of particle 2 with particle 3. For atoms the energy to remove the most loosely bound electron, i.e. the first ionization potential (Fig. 7.6(a), p. 131), is in the range ~ 2 eV to ~ 25 eV (e.g. it is 13.6 eV for hydrogen). Now the mass equivalent of 13.6 eV is 2.4×10^{-35} kg, or 2.6×10^{-5} times the electron rest mass, and it is clear that such small changes in mass cannot be detected by any normal weighing techniques. The very latest methods of mass spectrometry, developed from techniques similar to those of Thomson, can just detect the mass equivalents of electron binding energies. The energy required to remove a proton or a neutron (Fig. 7.6(b)) from a nucleus is approximately one million times the atomic electron binding energies, and its measurement is relatively easy. Such nuclear binding energies are directly involved in the production of power from nuclei, whether by fission or fusion, and it is not surprising that Einstein's equation $E = mc^2$ is so frequently associated in many people's minds with nuclear physics.

Problems

2.1. Marie and Pierre Curie remarked that for their first sample of polonium the 'activity is about 400 times as great as that of uranium', and for radium it is even greater. Comment on these statements in the light of later measurements of the half-lives: $\tau_{\frac{1}{2}}(^{238}\text{U}) = 4.51 \times 10^9$ years, $\tau_{\frac{1}{2}}(^{210}\text{Po}) = 140$ days, $\tau_{\frac{1}{2}}(^{226}\text{Ra}) = 1620$ years.

2.2. How 'complete' was the recovery of Becquerel's 'deactivated' uranium after his eighteen-month wait? The half-life of uranium X may be taken as 24.1 days.

2.3. Explain in words the expression for the mean life τ of a radioactive nucleus in a sample of N_0 such nuclei of radioactive constant λ,

$$\tau = \frac{1}{N_0} \int_0^\infty t(\lambda N . \mathrm{d}t).$$

Hence substitute $N = N_0 \exp(-\lambda t)$ and integrate by parts to show that $\tau = 1/\lambda$; see Fig. 2.1. (Lifetimes of unstable elementary particles, discussed in Chapters 7, 8 and 9, are expressed as 'mean life', τ, rather than 'half-life', $\tau_{\frac{1}{2}}$. See Table 8.1, p. 148.)

2.4. Study Fig. 2.2 and discuss expected observable differences between the simple thorium–thorium X scheme proposed by Rutherford and the series of decays now known to take place. (Hint: the thorium was in a thin layer, and a steady flow of air was maintained through the electroscope and the container.)

2.5. One gram of radium in radioactive equilibrium with its decay products emits 1.3×10^{11} alpha-particles per second. (a) Calculate the volume of helium gas at S.T.P. produced in a year. (b) Sir James Dewar (1908) used 70 mg of radium chloride ($RaCl_2$) in radioactive equilibrium and detected helium after waiting 1000 hours. What volume of gas at S.T.P. did he manage to detect? (Atomic weights: Ra 226, Cl 35.5.)

2.6. The present day abundancies of ^{238}U and ^{235}U are in the ratio $137.8:1$. Estimate the age of the Earth, assuming that the abundance ratio was originally $1:1$.

$\tau_{\frac{1}{2}}(^{238}U) = 4.5 \times 10^9$ years, $\tau_{\frac{1}{2}}(^{235}U) = 7.13 \times 10^8$ years.

2.7. The radioactive constants for the members of the radioactive series $A \to B \to C \to D$ are 4λ, 3λ, 2λ, and λ respectively, and the decay product of D is stable. Show (a) that the total activity of the sample decreases as $\exp(-\lambda t)$, and (b) that this condition is independent of the initial proportions of A, B, C, and D.

2.8. (a) A mixed beam of protons, deuterons, and alpha-particles accelerated from rest by a potential difference of 90 kV moves perpendicular to a uniform magnetic field of flux density 1.0 T. Calculate the radii of the paths of the three types of particles. (Assume ratios of masses to be $1:2:4$.) (b) What would happen if the same beam entered an electric field of 500 kV m^{-1}, in a direction perpendicular to this field, between parallel electrodes bent to form concentric semicircles of radii such that the path of the proton beam is also a (concentric) semicircle?

2.9. In J. J. Thomson's positive-ray apparatus there are five design parameters to be selected X, B, l, L and the accelerating potential difference on the gaseous discharge tube. Suggest suitable values and plot the expected parabolas for singly charged neon ions of mass numbers 20 and 22.

2.10. At what value of $T/m_0 c^2$ will the classical formula for kinetic energy $T = \frac{1}{2} m_0 v^2$ be in error by 10 per cent compared with the relativistic formula? What is the corresponding value of T for electrons? Does the classical formula give a value of $T/m_0 c^2$ that is too large or too small?

3. The nucleus revealed by alpha-particles

Introduction

The discoveries of one generation often become the tools of the next. There was a very short interval between the discovery of X-rays and their application in medical diagnosis and research, the result, in part, of the stimulation of popular imagination by newspaper reports. In similar fashion, the medical aspects of radioactivity were investigated soon after Becquerel in 1901 suffered skin burns as a consequence of carrying a radium source in his pocket for a few days. Within physics and chemistry the use of X-ray diffraction in the determination of crystal and molecular structures has developed steadily from the early work of W. H. Bragg right up to the automated and computerized systems available today. Electron beams at low energies were used by Franck and Hertz (1914) to investigate the quantized energy states of atoms, and the development of gas discharge tubes, and subsequently thermionic vacuum tubes, provided the first devices for electronics in science and industry. Applications of beta-particles and gamma-radiation in physics research were at first less spectacular, but the role of alpha-particles was of major significance.

To investigate an object, whether micro- or macroscopic, there are basically two available methods—to analyse the information spontaneously or systematically emitted by the object, or to interact with the object under controlled conditions (e.g. to bombard it) and observe the effects produced, including disintegration. The study of radioactivity is a good example of the former and the use of alpha-particles to bombard atoms is, even today, an important example of the latter.

We can distinguish three types of processes arising from the bombardment of an isolated object or target A by an incident particle a (Fig. 3.1):

1. *Elastic scattering.* The two particles retain their identities and their internal energies are unchanged. There is, of course, conservation

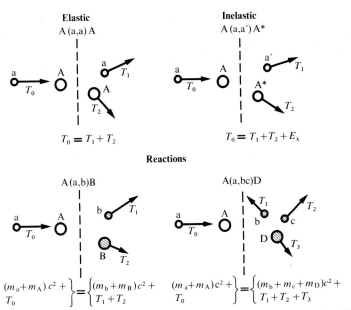

Fig. 3.1. Schematic representation of main types of nuclear interaction processes. The target A is initially at rest. T is the kinetic energy of a particle. For the inelastic scattering, E_x is the excitation energy of A*. In the energy conservation for the reactions m is the nuclear mass. Linear, and angular momentum are also conserved in all the processes.

of energy in the process, and in this case total kinetic energy T is conserved.

2. *Inelastic scattering*. The two particles retain their identities, but either or both of their internal energies are changed. The change E_x in the sum of their internal energies, by excitation or de-excitation, is equal to the change in the total kinetic energy.

3. *Reactions*. The identities of the resultant particles are different from those of the initial particles. One initial particle but not both may retain its identity, with or without a change of internal energy.

Energy conservation must be linked with changes in total mass by Einstein's relation $\Delta E = \Delta mc^2$, as indicated in Fig. 3.1.

Alpha-particle scattering

As early as 1906 Rutherford noticed effects due to small-angle scattering of alpha-particles in their passage through a thin foil of metal. Later, with Geiger, 'attention was directed to a notable scattering of the alpha-particles in passing through matter', and the following year (1909) Geiger and Marsden reported more detailed studies. They used thick reflectors (thicker than the range of the alphas) of eight different metals ranging from aluminium to lead, and found that the number of alphas scattered at an angle of about 90° increased with atomic weight A. As a result of work with thin foils they comment that 'it seems surprising that some of the alpha-particles, as the experiment shows, can be turned within a layer of 6×10^{-5} cm of gold through an angle of 90°, and even more. To produce a similar effect by a magnetic field, the enormous field of 10^9 absolute units [10^5 tesla] would be required.' In an estimate of the probability of scatter into backward angles by a thick plate of platinum, they made the simple assumption based on their observations that the scattered alphas 'were distributed uniformly round a half sphere', and found that 'about 1 in 8000 was reflected under the prescribed conditions'. The restrained comment of Geiger and Marsden ('it seems surprising') was evidently not an adequate description of Rutherford's reaction for in 1936 he declared in his last public lecture, 'it was quite the most incredible event that has ever happened to me in my life. It was almost as incredible as if you had fired a 15-inch shell at a piece of tissue paper and it came back and hit you.' But why was it so surprising?

At the time of the experiments Rutherford accepted J. J. Thomson's theory of the atom as the best working model for estimates of the scattering of both alpha- and beta-particles. 'The atom is supposed to consist of a number N of negatively charged corpuscles, accompanied by an equal quantity of positive electricity uniformly distributed throughout a sphere.' This quotation is from Rutherford's classic paper on 'The scattering of α and β particles by matter and the structure of the atom' (Rutherford 1911), and it should be noted that the apparent agreement between Crowther's measurements of beta scattering and Thomson's theory (1910) was already known to Rutherford. We can reconstruct Rutherford's reasoning using the expressions derived by Thomson.

The average deflection φ_1 due to a single sphere of radius R of positive electricity, is given by

$$\varphi_1 = \left(\frac{1}{4\pi\varepsilon_0}\right)\frac{\pi}{4}\frac{NeE}{Mv^2}\frac{1}{R},$$

where M, v, and E are the mass, velocity, and charge of the incident particle. The average deflection φ_2 due to the N electrons in the atom is given by

$$\varphi_2 = \frac{64}{5\pi}\left(\frac{3}{2N}\right)^{\frac{1}{2}}\varphi_1 = \frac{5.0}{N^{\frac{1}{2}}}\varphi_1.$$

Since the total mean deflection $\varphi = (\varphi_1^2 + \varphi_2^2)^{1/2} = \varphi_1(1 + 25.0/N)^{1/2}$, the electrons contribute, on Thomson's theory, about 15 per cent of the mean deflection for $N = 79$ (gold).

The mean-square scattering angle θ_x^2 for a foil of thickness x is φ^2 times the number of atomic collisions. In a foil with n atoms per unit volume, each of area πR^2, the number of encounters is $nx\pi R^2$, hence $\theta_x = (nx\pi R^2)^{1/2}\varphi$. For gold $n = 5.9 \times 10^{28}$ m^{-3}, and for $x = 2.5 \times 10^{-7}$ m, with $R = 6 \times 10^{-10}$ m, the value of θ_x for 5.5 MeV alpha-particles is $3.75°$. Geiger (1908) obtained an experimental value of about $3°$. However, as Rutherford noted, 'A simple calculation based on the theory of probability shows that the chance of an alpha-particle being deflected through $90°$ is vanishingly small.' For compound or multiple scattering to produce a deflection at angles greater than θ the probability is $\exp(-\theta^2/\theta_x^2)$, and if θ is set at $90°$ with $\theta_x = 3°$ the probability is about $e^{-900} \simeq 10^{-400}$. The cause of Rutherford's surprise is evident.

It was clear to Rutherford that larger scattering angles would be more probable as a result of single scattering if the radius R of the positively charged sphere were smaller. He therefore considered the limiting case of the charge concentrated at a point. In order to do justice to Rutherford's theory it is convenient to introduce some of the concepts involved and show their relation to measurements and known physical properties.

Differential cross-section

Scattering experiments and their theoretical interpretation are very important in many branches of physics. The probability of scattering is nowadays most commonly expressed in terms of the 'differential cross-section' and the utility and significance of this quantity are best introduced in terms of experimental arrangements and observables, i.e. on an operational basis.

Consider a collimated beam of N_0 particles per second incident on a target of thickness x (Fig. 3.2). A detector subtending a small solid angle $\delta\Omega$ at the region of incidence on the target receives N_θ particles per second scattered at a mean angle θ. For thin targets the scattering can be assumed to arise from single deflections, and it is then obvious that N_θ is proportional to N_0, and also to $\delta\Omega$, and also to the number of nuclei in a given area of the target. For reasons discussed later we use the number nx per unit area, where n is the number per unit volume, i.e. N_θ is proportional to $N_0 nx\,\delta\Omega$. The coefficient of proportionality is the differential cross-section, usually written $d\sigma(\theta)/d\Omega$, or more precisely $d\sigma_{el}(\theta)/d\Omega$ for elastic scattering,

hence
$$N_\theta = \frac{d\sigma(\theta)}{d\Omega}\, N_0\, nx\, \delta\Omega.$$

The interpretation of the words 'differential cross-section' now becomes clear by rearranging the terms

$$\frac{d\sigma(\theta)}{d\Omega} = \frac{N_\theta}{N_0}\frac{1}{nx}\frac{1}{\delta\Omega}.$$

We may define $d\sigma(\theta)/d\Omega$ as the probability N_θ/N_0 of scattering at an angle θ into unit solid angle by a target with one atom per unit area. The practical impossibility of a target with one atom per unit area serves to reveal the significance of the phrase 'cross-section', since the probability of the scattering event is the fraction of any given area of the target which is blocked out by the 'effective cross-sectional area' of

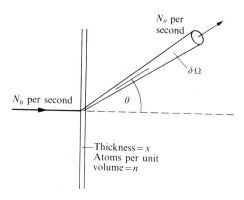

Fig. 3.2. Basic parameters in a nuclear scattering experiment.

the nuclei in that area. By choosing unit area of the target for nx and setting $nx = 1$ we obtain a definition of $d\sigma/d\Omega$ that relates directly to the effective cross-section for one nucleus. The units of $d\sigma/d\Omega$ are therefore $m^2 \, sr^{-1}$. A common alternative is $b \, sr^{-1}$ where $1b = 1$ barn $= 10^{-28} \, m^2$ and is of the order of the 'geometric' cross-section of the matter distribution in a nucleus of mass number $A \simeq 90$.

Note that the 'effective cross-sectional area' for scattering at a given angle as defined in the differential cross-section $d\sigma(\theta)/d\Omega$ relates to the probability of scattering into unit solid angle and there are 4π such units, steradians, available for any scattering in three-dimensional space. The word 'differential' implies that $d\sigma(\theta)/d\Omega$ is not the whole of the cross-section and to obtain the corresponding 'total cross-section' we need to integrate over the whole 4π of solid angle. The 'total elastic cross-section' is, therefore,

$$\sigma_{el} = \int \left(\frac{d\sigma(\theta)}{d\Omega} \right) d\Omega = \int_0^\pi \left(\frac{d\sigma(\theta)}{d\Omega} \right) 2\pi \sin \theta d\theta.$$

Thus for a spherically symmetric distribution of scattered particles, as encountered, for example, for low-energy (10–10^4 eV) neutrons elastically scattered by protons (see pp. 123 ff), $\sigma_{el} = 4\pi d\sigma(\theta)/d\Omega$. Such spherically symmetric distributions of scattered particles are confined to low-energy events, and normally the elastic differential cross-section has considerable angle dependence (see Figs 3.4 and 3.6). For charged particles there is also the problem that (simple) theoretical predictions of $d\sigma_{el}(\theta)/d\Omega$ tend to infinity as θ tends to zero. In practice atomic electron shielding of the nuclear charge becomes significant and invalidates such formulae at small angles.

But it is also evident that there are other cross-sections associated with the various possible inelastic scatterings and also the allowable reactions, and these will be large or small according to the probability of the process. Thus ^{238}U has values of σ_{abs} between 10^3 b and 10^4 b for the absorption of neutrons with energies within narrow bands (resonances) between about 6 eV and 100 eV, but the average σ_{abs} in this energy range is about 10 b (Fig. 6.2). At the other extreme we consider the minute total cross-sections for neutrino interactions. For example, the neutrinos associated with beta-decay have cross-sections of the order of 6×10^{-20} b for interactions with protons, a reaction that produces neutrons and positrons (see pp. 140 ff).

We may summarize the factors that determine the magnitude of $d\sigma(\theta)/d\Omega$.

1. Scattering angle θ—for Rutherford or Coulomb scattering the function is $\mathrm{cosec}^4(\theta/2)$.

2. Nature of incident particle, notably its mass and electric charge, but also a number of other factors such as its intrinsic angular momentum or spin, and whether these spins are preferentially aligned (i.e. polarized) in relation to the scattering plane.

3. Kinetic energy T_1 of incident particle—for Rutherford scattering the function is $1/T_1^2$.

4. Nature of the target nucleus—similar factors as for the incident particle, including the possibility of aligned spins.

5. Kinetic energy of target nucleus—normally taken as zero except in colliding beam experiments (see p. 68 ff).

6. Type of process: (a) elastic scattering; (b) inelastic scattering (specify excitation energy state(s) involved); (c) reaction (specify in detail).

Rutherford's theory

The only type of process treated at this time by Rutherford was elastic scattering. The simplest possible case is for 180° scattering (Fig. 3.3) and this introduces a useful quantity b_0, the distance of closest approach for a head-on collision. At the point of closest approach the kinetic energy T_1 has been changed into electrostatic potential energy

$$T_1 = \frac{1}{4\pi\varepsilon_0} \cdot \frac{q_1 q_2}{b_0}$$

hence

$$b_0 = \frac{1}{4\pi\varepsilon_0} \cdot \frac{q_1 q_2}{T_1}.$$

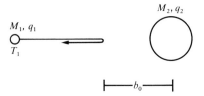

Fig. 3.3. Distance of closest approach, b_0, for 180° scattering.

Rutherford obtained an estimate of b_0 using $q_2 = 100e$ and $q_1 = 2e$. For gold, q_2 is now known to be $79e$, and for alphas $q_1 = 2e$. For $T_1 = 5.6$ MeV, $b_0 = 40.6 \times 10^{-15}$ m. (10^{-15} m = 1 fm, i.e. one femtometer, but is commonly termed a fermi, after the famous Italian-American nuclear physicist.) Since the radius of the atom is about 10^{-10} m it is clear that the alphas penetrate so deep into the electron distribution that the only effective electric field acting on them is that of the nucleus, at least for scattering angles greater than about 1°.

Rutherford derived an expression for the differential cross-section for elastic scattering at an angle θ on the basis of the following assumptions: (1) the only forces acting are electrostatic; (2) the incident particle does not penetrate the positive charge distribution of the target atom; (3) the electrons in the atom have negligible effect; (4) the recoil of the heavy target nucleus is negligible; (5) the target is thin enough for one target nucleus not to obscure another and for multiple scattering and energy losses to be neglected. The derivation is given in Appendix A, and the result is easy to remember:

$$\frac{d\sigma(\theta)}{d\Omega} = \frac{b_0^2}{16} \operatorname{cosec}^4(\theta/2),$$

where b_0 is the distance of closest approach for $\theta = 180°$. For a detector of area D at a distance r from the target,

$$N_\theta = \frac{N_0}{16} \cdot \frac{D}{r^2} \cdot \frac{N_A \rho x}{A} \cdot \left(\frac{q_1 q_2}{T_1}\right)^2 \frac{\operatorname{cosec}^4(\theta/2)}{(4\pi\varepsilon_0)^2},$$

since $n = N_A \rho / A$, where N_A is the Avogadro constant and ρ is the density of the target of atomic weight A.

In order to impress on the reader the orders of magnitude involved and their dependence on θ, the magnitude of $d\sigma(\theta)/d\Omega$ is shown in Fig. 3.4 for $q_1 = 2e$ (alpha), $q_2 = 79e$ (gold), $T_1 = 5.6$ MeV. The corresponding N_θ is also shown for the further conditions $N_0 = 10^6$ s^{-1}, $D = 1$ cm^2, $r = 0.3$ m, ρ(gold) $= 19\,300$ kg m^{-3}, A(gold) $= 197$, $x = 10^{-6}$ m. Note that the vertical scales are logarithmic.

Rutherford reported in the same paper that Geiger had found that measurements between 30° and 150° for gold were in agreement with the theoretical angular distribution. Later Geiger and Marsden (1913) confirmed this between 5° and 150° and also showed that N_θ was proportional to the thickness x and to the inverse square of the kinetic energy T_1. They counted over 100 000 scintillations produced by

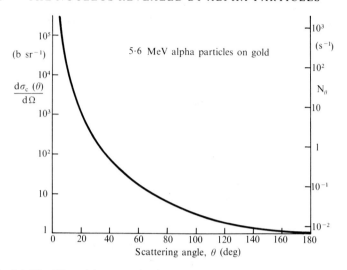

Fig. 3.4. The differential cross-section for Rutherford scattering of 5.6 MeV alphas by gold, and the corresponding count rates for 10^6 alphas per second incident on a gold foil of thickness 10^{-6} m with a detector of area 1 cm^2 placed 30 cm from the target. Notice the logarithmic scale.

individual alpha-particles incident on a detector of zinc sulphide (see pp. 81 ff). The relation to q_2 was more difficult to determine as the absolute numbers of scattered particles had to be measured. Nevertheless, the nuclear charge number q_2/e for gold was estimated to be within 20 per cent of half the atomic weight; the accepted values are $q_2/e = 79$, $A = 197$.

The conclusions to be drawn from this work are as follows.

1. The major part of an atom (i.e. all except the relatively light electrons) is concentrated in a nucleus with a positive electric charge equal to approximately $(A/2)e$.

2. The interaction between incident alpha-particles of energies between about 2 MeV and 6 MeV and nuclei ranging from aluminium to gold can be adequately explained by the electrostatic force alone, and the alpha particle at its point of closest approach has not penetrated the charge distribution of the target nucleus. We may interpret this as setting an upper limit for the radius of the nucleus $R < b_0 = 40.6 \times 10^{-15}$ m for gold, and $R < b_0 = 6.7 \times 10^{-15}$ m for aluminium.

Rutherford's theoretical interpretation of Geiger and Marsden's experiments firmly established the picture of the nuclear atom, thereby paving the way for theories of atomic structure and detailed studies of the nucleus itself.

Diffraction effects in alpha-scattering

The next major development in the study of the elastic scattering of heavy charged particles (i.e. alpha-particles and, later, protons, etc.) came with the construction of nuclear accelerators, in particular the simple cyclotron (see pp. 62 ff). Incident particles of higher energies approach more closely to the nucleus, and at a certain distance the Rutherford formula begins to be inaccurate. This occurs at a particular energy for observations at a fixed angle (Fig. 3.5(a)), and correspondingly for an adequately high energy becomes evident at a particular angle (Fig. 3.5(b)). The effects are more noticeable for lighter atoms with their lower nuclear charge Ze. At a high enough energy an important new phenomenon is found for nearly all incident particles and target nuclei, namely, an angular distribution characteristic of diffraction effects (Fig. 3.6). The interpretation in terms of the

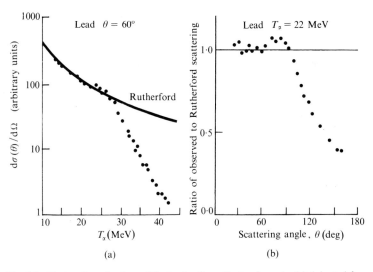

Fig. 3.5. Limits of application of Rutherford's scattering formula. (a) Adapted from G. W. Farwell and H. E. Wegner, *Phys. Rev.* **95**, 1212 (1954). (b) Adapted from N. S. Wall, J. R. Rees, and K. W. Ford, *Phys. Rev.* **97**, 726 (1955).

wave nature of moving particles, using the de Broglie wavelength $\lambda = h/Mv$, yields further information on the size of the nucleus. Thus for the simplest possible model, in which the nucleus is treated as a black disc of radius R the angle of the first diffraction minimum is $\theta_{min} \simeq \sin^{-1} (0.61\lambda/R)$. Hence for $\lambda = R$, $\theta_{min} \simeq 38°$. Since $\lambda_\alpha = h(2M_\alpha T_\alpha)^{-1/2}$, $\lambda_\alpha = 4.0$ fm at $T_\alpha = 18$ MeV, suggesting $R \simeq 5.5$ fm for ^{40}Ca (Fig. 3.6) for which $\theta_{min} \simeq 26°$. A more sophisticated treatment is provided by the optical model of the nucleus (see, *The properties of nuclei* by G. A. Jones, OPS 12) which allows for reflection, refraction, and absorption ('cloudy crystal ball') and introduces a surface diffuseness in place of a hard, abrupt surface. Its ability to account for both elastic and inelastic scattering, and for polarization effects (see, p. 132), is well established for both charged and uncharged particles. The required size and shape of the nuclear matter distribution is closely related to the electric charge distribution deduced from comparable experiments with high-energy electrons (T_e is a few hundred MeV) and the commonest simple function used to describe it is the so-called Fermi or Saxon–Woods formula,

$$\rho_r = \frac{\rho_0}{1 + \exp\{(r - R)/a\}},$$

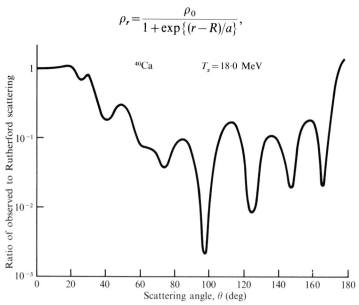

Fig. 3.6. Diffraction pattern observed for 18 MeV alpha-particles scattered by ^{40}Ca. (Adapted from C. P. Robinson, J. P. Aldridge, J. John, and R. H. Davis, *Phys. Rev.* **171**, 1241 (1968).)

where ρ is the matter or charge density. This is illustrated in Fig. 3.7 for calcium ($A = 40$) and gold ($A = 197$) using values for R obtained from the important relation deduced from a wide range and variety of experiments,

$$R = r_0 A^{\frac{1}{3}},$$

with $r_0 = 1.1$ fm. The diffuseness parameter a is set at 0.65 fm and it can be shown that the surface region from 10 per cent to 90 per cent maximum density extends over $4.4a \simeq 3$ fm. The values of r_0 and a for n, p, d, ^{3}H, ^{3}He, and alpha-scattering, for energies up to at least 100 MeV, do not need to be varied by more than about 15 per cent from the values used here. It is clear from the Fermi formula that R is the 'half-density' radius, and to a good approximation the central charge and matter densities and also the surface thickness are the same for all nuclei, excepting the very lightest.

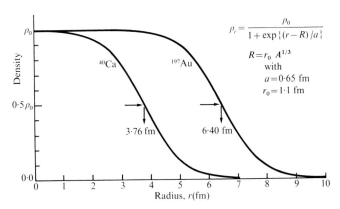

Fig. 3.7. Radial dependence of nuclear charge density given by Saxon-Woods (or Fermi) formula with parameters r_0 and a obtained from charged-particle scattering.

Alpha-particle spectra

The first indication that the alpha-particles emitted by a given species of nucleus were not all of the same energy was noted by Rutherford and Wood (1916). They found that about 1 in 10^4 alphas from a thorium C (^{212}Bi) source had a range of some 113 mm in air, (equivalent to $T_\alpha = 10.6$ MeV) and about 3 in 10^5 had a 100 mm range.

The principal group had a range of 86.2 mm (8.8 MeV). Comparable long-range alphas were also found for radium C (^{214}Bi).

As late as 1930, Rutherford, Chadwick, and Ellis, in their classic volume *Radiations from radioactive substances*, were unable to decide between two possible origins: (1) a 'separate radioactive product' with an exceedingly short decay period, as deduced from the Geiger–Nuttall relation, and (2) an extra supply of energy given to the normal alphas 'just before their ejection'. They were percipient enough to note that the extra energy was approximately equal to the energies of gamma-rays from radioactive nuclei.

Rosenblum (1929) used a powerful electromagnet to deflect a beam of alpha-particles through 180°, thereby achieving a focusing effect that facilitated accurate measurements of the radii of curvature of the tracks of the alphas, and hence their energies. He found not only the rare high-energy alphas but also groups of lower energies. In particular for thorium C (^{212}Bi) six groups are found at 6.086 MeV (27.2 per cent), 6.046 MeV (69.9 per cent), 5.758 MeV (1.7 per cent), 5.615 MeV (0.15 per cent), 5.603 MeV (1.0 per cent), and 5.478 MeV (0.014 per cent).

By a close analogy with the spectral lines from excited atoms, Ellis (1922) explained various relations between gamma-ray energies (e.g. $h\nu_1 + h\nu_2 = h\nu_3$) in terms of excited states of nuclei. It was later realized that the differences between the alpha-particle energies for thorium C corresponded to the observed energies of associated gamma-rays (e.g. 40 keV, 328 keV, 471 keV), and the careful construction of energy-level schemes finally led to the representation shown in Fig. 3.8. On this diagram the energies of the emitted alpha-particles are indicated, but in order to calculate accurately the corresponding energy levels and the consequent gamma-rays it is essential to allow for the recoil of the daughter nucleus. Thus for the 6.086 MeV alpha-particle the ^{208}Tl recoil energy is 117 keV, and for the 5.603 MeV alpha it is 108 keV, from $T_{208} = (4/208)T_4$.

The explanation of the long-range alphas now follows naturally— the nucleus (^{212}Po) is excited (e.g. by 1.81 MeV) when the alpha is emitted, and the total available energy is therefore that much greater. Most excited states of nuclei decay by gamma-emission in times of the order of 10^{-12} s. For ^{212}Po in the ground state the half-life for alpha-emission is 3×10^{-7} s and for the excited state the half-life is much shorter. There is thus competition between gamma-emission and alpha-emission. This explanation sits astride the two possibilities of

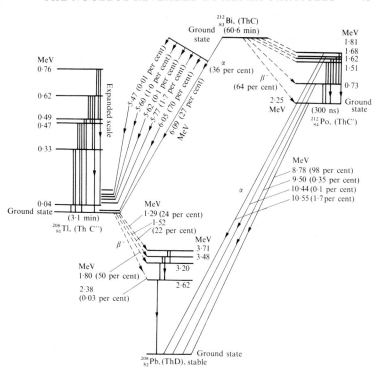

Fig. 3.8. Decay scheme for ^{212}Bi with alpha and beta energies, and the subsequent gamma energies, correlated in the energy levels of the daughter and parent nuclei.

Rutherford, Chadwick, and Ellis if we allow that ^{212}Po excited by 1.81 MeV is a 'separate radioactive product'.

Several features of radioactive decay processes are illustrated in Fig. 3.8.

1. Branching of a radioactive series—^{212}Bi decays 36 per cent by alpha-emission and 64 per cent by beta-emission.

2. Emission of betas of several maximum energies by the same nuclear species, with correlated gamma-rays from the excited daughter.

3. The stable end product ^{208}Pb. This is a nucleus with a number of remarkable properties. Note, for example, the very large value (2.62 MeV) of the energy of its first excited state compared with

0.040 MeV (^{208}Tl) and 0.726 MeV (^{212}Po). (^{208}Pb is 'doubly magic', $Z = 82$, $N = 126$; see pp. 130 ff.)

4. In balancing energy changes for the alternative routes between ^{212}Bi and ^{208}Pb it is essential to use maximum beta energies.

5. Some gamma transitions are 'forbidden', i.e. the probability of gamma emission between two particular energy states is too small to allow detection, e.g. the ^{208}Tl level at 0.492 MeV 'cannot' de-excite direct to the ground state. The gammas in time-coincidence with the 5.62 MeV alphas have energies 0.452 MeV and 0.040 MeV.

Alpha-emission

Geiger and Nuttall (1911) proposed a relation (see p. 22) between the energy of emission of alpha-particles T_α and the radioactive constant λ of the parent nucleus, $\ln \lambda = A_1 \ln T_\alpha + B_1$, where A_1 and B_1 are constants. The derivation of a closely similar relation, by Gurney and Condon (1928) and also by Gamow (1928), was one of the early triumphs of wave mechanics.

Consider the reverse of alpha-emission, i.e. the alpha-particle energy required to approach a nucleus of atomic number Z. This is simply

$$T_\alpha = (1/4\pi\varepsilon_0)zZe^2/r$$

where r is the distance from the centre of the alpha-particle, of charge ze, to the centre of the nucleus of charge Ze. If we assume that the alpha-particle on reaching the nuclear surface at $r = R$ is absorbed into and bound by the nucleus, then the simplest potential energy diagram is the repulsive Coulomb potential with a cut-off to a square well at $r = R$, as shown in Fig. 3.9. We need not, for the present discussion, attempt to specify $V(r)$ for r less than R. It is clear that, on classical grounds, no alpha-particle can escape from the potential 'well' unless its final kinetic energy a long way from the nucleus is greater than the 'peak' value of the potential. A few simple order of magnitude calculations will show that classical physics cannot explain alpha-emission. Thus, for $^{224}_{90}$Th alpha-emission the residual nucleus is $^{220}_{88}$Ra, for which $R = r_0 A^{1/3} = 6.64$ fm for $r_0 = 1.1$ fm, and for a maximum reasonable value of $r_0 = 1.5$ fm we have $R = 9.05$ fm. The corresponding peak energies are 38.2 MeV and 28.0 MeV, many times larger than the highest observed energies of emitted alpha-particles (~ 10 MeV). Even if the finite size of the alpha-particle is included, and the surface diffuseness of the nucleus is allowed for, we cannot take the effective radius for absorption to be greater than about 13 fm at which the peak energy becomes 19.5 MeV, still too large to allow 3–10 MeV alphas to escape by classical means. A supplementary difficulty for classical physics arises from the fact that nuclei emitting 4 MeV alphas do not detectably absorb 7 MeV alphas, although energetically this ought to be possible.

The recognition that alpha-particles have wave-like properties, and require the application of wave mechanics, leads directly to the result that there is a finite probability for alpha-particles within the nucleus to 'penetrate' or 'tunnel through' the barrier. A more realistic form of the potential $V(r)$ for $r > R$ is that obtained from alpha-scattering experiments (see p. 41) by Igo (1958) interpreted by the optical model of the nucleus (see pp. 42 ff), which is also shown in Fig. 3.9. Neither

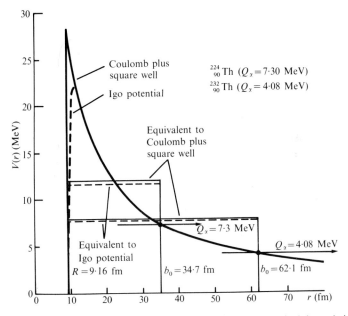

Fig. 3.9. Potential barriers for thorium used in the analysis of alpha-emission probabilities. The equivalent square barriers are calculated to give the same transmission probabilities as those of the realistic Igo potential and the 'Coulomb plus square-well' potential, and use barrier widths of b_0-R for the $Q_\alpha = 7.3$ MeV and 4.08 MeV, where b_0 is the distance of closest approach for 180° scattering by the daughter nucleus.

this form nor the Coulomb potential with a cut-off to a square well is simple enough to encourage an analytic solution of the wave equation. To appreciate the physics of the process, it is most easy to consider a one-dimensional square barrier of width b (Fig. 3.9), with an incident beam of alpha-particles of energy T_α that is less than the height of the barrier V. The wave equation to be solved is then

$$d^2\psi/dr^2 + [2\mu_\alpha\{Q_\alpha - V(r)\}/\hbar^2]\psi = 0,$$

where $\mu_\alpha = M_\alpha(A-4)/A$ is the reduced mass of the alpha-particle emitted by a nucleus of mass number A, and Q_α is the total available energy of the alpha and the nuclear recoil, $Q_\alpha = T_\alpha A/(A-4)$. Notice the simplifying assumption that $V(x) = 0$ for $x < -b$, corresponding to inside the nucleus, where the form of the potential is not at all well known. The solution of this wave equation is derived in Appendix B and indicates that the probability of transmission through a square barrier of height V and width b is given by

$$\text{transmission} = 4(Q_\alpha/V)(1 - Q_\alpha/V)(1 + \sinh^2\beta b)^{-1}$$

where Q_α is the total available alpha-particle energy (including recoil) and β is given by

$$\beta^2 = 2\mu_\alpha(V - Q_\alpha)/\hbar^2.$$

In order to apply this result to alpha-emission we need to make further simplifying assumptions. We first note that the three-dimensional solution $\psi(r)$ for a spherically symmetric nucleus, emitting an alpha-particle along a radius, can be written $(1/r)U(r)$, where $U(r)$ turns out to be identical in form with $\psi(x)$, and we therefore simply substitute r for x and U for ψ. A glance at Fig. 3.9 suggests that suitable values for V and b of the equivalent square-barrier potential are about 15 MeV and 30 fm respectively for Q_α = 7.30 MeV, and for these values $\beta b = 36$. It follows that for such large values of βb a very good approximation is

$$(1 + \sinh^2\beta b)^{-1} = 4\exp(-2\beta b),$$

and the transmission is then $16(Q_\alpha/V)(1 - Q_\alpha/V)\exp(-2\beta b)$.

The radioactive constant λ is the probability of transmission multiplied by the frequency of impact of alphas on the barrier. We shall assume that this impact frequency is not sensitively dependent on the alpha-emission energy Q_α. We note that the factor $16(Q_\alpha/V)(1 - Q_\alpha/V)$ is also not very sensitive to Q_α, e.g. it has a maximum value of 4 at $Q_\alpha = V/2$ and equals 3 at $Q_\alpha = V/4$ and $3V/4$. On the other hand, $\exp(-2\beta b)$ is markedly dependent on Q_α for a given V. Thus $\exp(-2\beta b)$ for $^{224}_{90}\text{Th}$ with $b = 25$ fm is 1.48×10^{-39} for $Q_\alpha = V/2$ $= 6$ MeV, but it is 1.22×10^{-55} for $Q_\alpha = V/4 = 3$ MeV, and 1.98×10^{-32} for $Q_\alpha = 3V/4 = 9$ MeV. In ratios, as Q_α is increased from 3 MeV to 6 MeV to 9 MeV the probabilities of emission increase as 10^{-16} to 1 to 10^7. In other words, for a square barrier of width 25 fm and height 12 MeV, the lifetimes decrease by a factor of 10^{23} as Q_α increases from 3 MeV to 9 MeV.

This extreme sensitivity is implicit in the Geiger–Nuttall relation (see p. 22),

$$\ln \lambda = A_1 \ln T_\alpha + B_1 \simeq A_1 \ln Q_\alpha + B_1,$$

with $A_1 = 82.5$ and $B_1 = -157$, for λ in s^{-1} and T_α in MeV, as deduced from observed values of Q_α for the thorium series with mass number $A = 4n$. This relation yields for $T_\alpha = 3$, 6 and 9 MeV the ratios 1.5×10^{-25} to 1 to 3.4×10^{14}, indicating an even greater sensitivity of lifetimes on alpha-energy than that predicted by the square barrier. Results from calculations with more realistic shapes for the potential barrier agree more closely with experimental values.

For the square barrier, the theoretical equivalent of the Geiger–Nuttall relation is shown in Appendix B to be

$$\ln \lambda = A_2 Z Q_\alpha^{-\frac{1}{2}} + B_2.$$

However, for the ranges of Q_α involved, the form of $Q_\alpha^{-1/2}$ is quite close to that for the $\ln Q_\alpha$ of the Geiger–Nuttall relation, as shown in Fig. 3.10. This figure also includes the form of the exact solution for the Coulomb potential with the square-well cut-off at radius R (Fig. 3.9), for which the relation is (see p. 187 for constant K)

$$\ln \lambda = -K Q_\alpha^{-\frac{1}{2}} \{\cos^{-1}(R/b_0)^{\frac{1}{2}} - (R/b_0 - R^2/b_0^2)^{\frac{1}{2}}\} + B_3.$$

The dependence is not now on $\ln Q_\alpha$ or on $Q_\alpha^{-1/2}$, but on $F Q_\alpha^{-1/2}$ with F also slightly dependent on Q_α through the terms in b_0. These three functional forms, Fig. 3.10, provide almost equally good overall fits to the experimental data.

Notice that the Coulomb plus square-well formula $F Q_\alpha^{-1/2}$ depends also on the nuclear radius R and allows estimates of this parameter which yield $R \simeq 1.5 \, A^{-1/3}$ fm, some 25 per cent larger than presently accepted values.

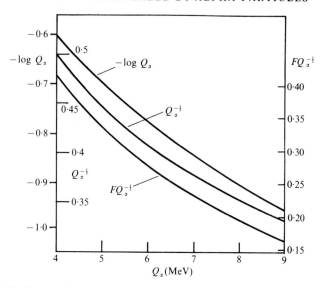

Fig. 3.10. The three functions encountered in versions of the Geiger–Nuttall relation, showing their close similarity for $Q_\alpha = 4 - 9$ MeV. Log Q_α is from the original empirical relation, $Q_\alpha^{-1/2}$ from the simple square-barrier theory and $FQ_\alpha^{-1/2}$ from the Coulomb potential with square cut-off. (Each of the three relations includes an adjustable constant that allows the arbitrary vertical displacement used to prevent confusing overlaps of the curves.)

The probability of alpha-emission is directly related to the probability for the inverse process, alpha-absorption. The same theory also applies to the emission and absorption of other charged particles, e.g. protons or $^7_3$Li nuclei, with suitable changes of parameters. For example, if it were energetically allowed the emission of $^7_3$Li with the same energy (4 MeV) and the same residual nucleus, which will be $^{288}_{88}$Ra, would have a half-life of about 4×10^{60} years, indicating an unobservably low level of activity. (This value is obtained for the truncated Coulomb potential.) The similar calculation for the emission of a 4 MeV proton, leading to the same residual nucleus, yields a half-life of 1.6×10^{-15} s. This is of the same order of magnitude as the lifetime of the compound nucleus formed when an incident particle coalesces with a target nucleus. Notice that so-called 'proton radioactivity' occurs when beta decay leads to a proton-emitting nucleus, and can have (beta-decay) lifetimes much larger than 10^{-15} s. By analogy to 'delayed neutron emission' (see p. 107) it is better to call this '(beta) delayed proton emission'.

Alpha-particle reactions

One of the problems solved by the wave-mechanical interpretation of alpha-emission (1928) was how it came about that 5 MeV alphas (say) were observed to be emitted by a given nucleus but even 7 MeV alphas were not observed to be absorbed by that nucleus or by nuclei of comparable A and Z. Barrier penetration theory indicates that for heavy nuclei the probability of exit (or entry) of a 7 MeV alpha is about 1 in 10^{22}, (see pp. 186 ff) but experiments to measure such small absorptions are notoriously difficult to undertake.

However, in 1919, long before wave mechanics, the effects of alpha-particle absorption reactions had been observed and correctly interpreted by Rutherford. Marsden (1914) noted in the course of scattering experiments that alpha-particles incident on hydrogen gas produced long-range particles. The particles passed through a foil that was thick enough to stop the alphas and then produced scintillations on a zinc-sulphide screen. They were considered to be protons, i.e. knock-on hydrogen nuclei from elastic collisions. Rutherford (1919) repeated the experiments with various gases and reported 'On introducing oxygen or carbon dioxide into the vessel, the number of scintillations fell off in amount corresponding with the stopping power of the column of gas. An unexpected effect was, however, noticed on introducing dried air . . . the number of scintillations increased . . . which appeared of about the same brightness as [hydrogen]-scintillations.' After further investigations he boldly surmised 'it is difficult to avoid the conclusion . . . that the nitrogen atom is disintegrated under the intense forces developed in a close collision with an alpha particle.' The closeness is indicated by the formula for the distance of closest approach, $b_0 = zZe^2/4\pi\varepsilon_0 T_\alpha$, where Z is small for the light elements ($b_0 = 3.5$ fm for $T_\alpha = 5.6$ MeV on nitrogen). All the elements from boron to potassium, apart from carbon, oxygen, and, possibly, beryllium, emitted protons under alpha bombardment.

Proposed interpretations included (1) the alpha knocks out a proton from the nucleus and itself remains an alpha; (2) the alpha is absorbed and a proton is emitted; (3) the alpha is broken up and the nucleus unchanged; and (4) both alpha and nucleus are broken up. Blackett (1925) used a cloud chamber (see pp. 83 ff) to make visible the tracks of the charged particles involved including the recoil nucleus, and showed unambiguously from 8 examples out of 400 000 alpha collisions that (2) is the most probable. We may write the nuclear reaction equation

$$^{14}_{7}N + {}^{4}_{2}He \rightarrow ({}^{18}_{9}F^{*}) \rightarrow {}^{1}_{1}H + {}^{17}_{8}O,$$

taking care to balance the electric charges of the nuclei involved and select the obvious mass numbers. The intermediate step, in which an excited 'compound nucleus' is formed (*indicates excitation), has been found to take place in many such reactions. The overall reaction is sometimes summarized as $^{14}N(\alpha, p)^{17}O$, and from the conservation of mass-energy we can deduce the energy of the proton for a given alpha-particle energy (see pp. 95 ff) if we know sufficiently accurately the masses of the nuclei involved.

The significance of Rutherford's observations was not under-estimated. This artificial transmutation of the elements had been claimed without adequate foundation not only by alchemists but also by nineteenth-century scientists. The wave-mechanical interpretation of alpha-emission indicated how very much more probable would be the entry of a proton into the nucleus, and this stimulated the building of (proton) accelerators. The first proton reaction was observed with ^{7}Li by Cockcroft and Walton (1932) using 125 keV protons.

The important role of alpha-particle reactions was not yet eclipsed, for in the same year Chadwick introduced the neutron to explain some highly penetrating radiation observed by Bothe and Becker (1930) when beryllium was exposed to 5.3 MeV alphas from polonium. The neutron was discussed as a possibility by Rutherford and others as early as 1920 and even systematically sought by Chadwick before 1930. In 1932 Irene Joliot Curie and Frederick Joliot discovered that Bothe and Becker's 'radiation' ejected protons of high velocity from a sheet of hydrogenous material, e.g. paraffin wax. The conservation laws of energy and momentum excluded the possibility that the radiation consisted of very high-energy gamma photons (see problem 6.13), and Chadwick proposed 'particles of mass nearly equal to that of the proton and with no net charge'. The neutron's lack of electric charge prevents it from producing ionization along its track, and explains why it was relatively difficult to detect and identify. It also suggests that it should react readily with nuclei since there would be no Coulomb potential barrier, and this was found to be the case (see pp. 104 ff).

Somewhat later it was found that the neutron is radioactive (Snell and Miller 1948) with a half-life now known to be $\tau_{\frac{1}{2}} = 633.4$ s. The decay process is

$$n \rightarrow p + e^- + \bar{\nu}_e + 0.780 \text{ MeV}$$

where $\bar{\nu}_e$ is the electron antineutrino (see pp. 143 ff).

Alpha-particle scattering led to the discovery of the nucleus, and alpha-particle reactions led to the discovery of the neutron, neutron reactions, and nuclear power generators. The discoveries of Rutherford's generation not only became the tools of the next generation but also provided one of its sources of power.

Problems

3.1. Check Geiger and Marsden's calculation (see p. 34) of the magnetic flux density required to turn alpha-particles through $\theta > 90°$ in a layer of thickness $d = 6 \times 10^{-7}$ m. List the assumptions and approximations made.

3.2. Use the formula $R = r_0 / A^{1/3}$ for the radius of a nucleus of mass number A to deduce the A number of the nucleus that has a 'geometric' cross-section of 1 barn. (Use $r_0 = 1.1$ fm.)

3.3. Calculate $d\sigma(\theta)/d\Omega$ for Coulomb scattering at $\theta = 90°$ for 5.6 MeV alpha-particles on gold, and hence the number scattered from an incident beam of 10^6 alphas per second into a detector of area 1 cm^2 placed 0.3 m from the target of thickness 10^{-6} m (Fig. 3.4).

3.4. A small source of alpha-particles, activity S, and a small semiconductor detector, area D, lie symmetrically on the axis of, and on opposite sides of, a narrow annular transmission target, radius R, width δR, thickness x, n nuclei per unit volume. Show that the count-rate for scattering at an angle θ is $(S\delta RnxDb_0^2/32R^3)\cos(\theta/2)$, where b_0 is the distance of closest approach of the alphas to the nucleus for 180° scattering.

3.5. In an experiment on Coulomb-scattering, replacement of a thin $^{64}_{30}$Zn target by a ^{114}Cd target of the same mass per unit area increased the scattered intensity by a factor 1.45. Calculate the atomic number of cadmium.

3.6. Calculate the recoil energy of (a) the ^{208}Tl nucleus when a 6.086 MeV alpha-particle is emitted from ^{212}Bi, and (b) the ^{137}Ba nucleus when a 660 keV gamma-photon is emitted. Use integral masses for (a) and M(Ba) = 137 u for (b).

4. Nuclear accelerators

Introduction

A great deal was discovered about the nucleus by the use of naturally occurring alpha-particles, but their maximum energy is only about 5.5 MeV for useful source strengths, and their double electric charge inhibits their near approach to all but the lightest nuclei. Machines were therefore designed to accelerate stable charged particles of any type, and the highest energies obtainable have become limited only by the cost associated with the consequent large size and complexity.

In practice, the experimental nuclear physicist is concerned with the characteristics of the beam of charged particles. These include (1) type(s) of particle, (2) energy and energy resolution, (3) spatial and angular distribution, (4) intensity, (5) time structure (see below), (6) polarization, and (7) stability of the factors (1) to (6). These characteristics are determined by the initial source of ions, the type of accelerator, and the beam transport system. The ion source usually involves a radio-frequency or electron-fed d.c. discharge and determines the type of particle, the intensity, and the polarization, i.e. the degree of alignment of the spins of the particles. The type of accelerator determines the energy and energy spread, the spatial and angular distribution, and the time structure (e.g. continuous current, pulses of particles, or pulses with a fine radio-frequency bunching). Beam transport is an important part of the process, especially in experiments with beams of unstable fundamental particles, (e.g. pions, kaons) which have to be produced by bombarding targets, sometimes within the accelerator, with very high-energy stable particles (e.g. protons). Selection of the desired type of particle is then achieved by magnetic and electric separators, based on principles that are similar to those of mass spectrometers. Together with magnetic deflectors, collimators, and a series of magnetic lenses, these particle separators also influence the band of energies selected, the transmitted intensity, and (within limits) the spatial and angular distribution.

There are two main types of accelerator. One type produces an accelerating electric field acting continuously on a bunch of particles

TABLE 4.1

	Particles	Maximum energy or voltage
1. Continuous acceleration		
(a) Cockcroft–Walton (voltage multiplier)	Any	$\simeq 5$ MV
(b) van de Graaff—single (electrostatic)	Any	$\simeq 20$ MV
van de Graaff—tandem (two stages with electron stripping half-way)	Start with negative ions	$\simeq 40$ MV
(c) Betatron (magnetic induction)	e	$\simeq 400$ MeV
(d) Electron LINAC (travelling electromagnetic wave in waveguide)	e	$\simeq 50$ GeV
2. Repeated acceleration using radio-frequency fields		
(a) Linear accelerator (resonant cavity with drift tubes)	Any	800 MeV p's
(b) Simple cyclotron (magnetic resonance)	Any	$\simeq 22$ MeV p's $\simeq 80$ MeV α's
(c) Sector-focusing cyclotron (fixed frequency)	Any	500 MeV p's
(d) Synchrocyclotron (frequency-modulated)	Any	1000 MeV p's ($\simeq 1$ GeV)
(e) Synchrotron—proton (vary B and possibly f)	Any	900 GeV p's ($= 1$ TeV) $\rightarrow$ 20 Tev?
(f) Synchrotron—electron (vary B)	e	100 GeV

(The availability of alternative machines with more desirable characteristics makes it unlikely that higher energy versions will be built of 1(a), 1(c), 2(b) and 2(d).)

or on a steady current of particles. The other type produces repeated accelerations in regions localized in space. The repetition implies alternating electric fields, and it is therefore necessary to shield the particles during intervals of unfavourable field directions or phases. A summary of the principal kinds of accelerator is given in Table 4.1, and the energies achieved since 1930 are shown in Fig. 4.1.

Continuous acceleration

The Cockcroft–Walton voltage-multiplier consists of rectifiers and capacitors (Fig. 4.2) and was the first accelerator (Cockcroft and

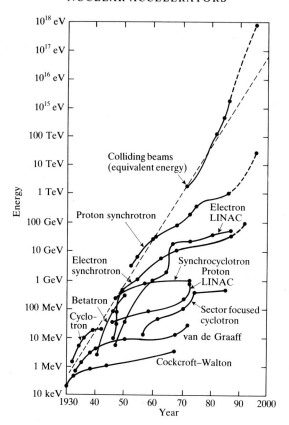

Fig. 4.1. Chronology of maximum beam energies from accelerators and equivalent energies for colliding beams.

Walton, 1930) used in nuclear-reaction studies. It is still often used as the first stage of even the largest accelerators. Its action is as follows: When B is positive with respect to A, it is clear that C_1 will tend to charge up, through the diode BC, to a maximum voltage V. If D is at a lower potential than that of C, then at the same time current will flow through diode CD and will tend to charge up C_2 and also provide current for diode DE, depending on the voltage already on C_2 and on the potential difference across DE. When B is negative with respect to A, the voltage acting across CB is eventually, after many cycles, a maximum of $+V$ from C_1 and $+V$ from the transformer secondary, i.e. C_2 will tend to charge up to $2V$ through diode CD, with diode BC

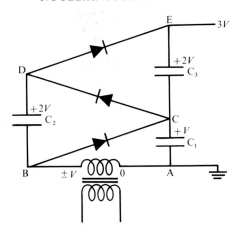

Fig. 4.2. Circuit diagram to illustrate the principle of the Cockcroft–Walton voltage multiplier.

not conducting because of the reversed polarity. Similarly, when B is positive with respect to A, the voltage acting across DC is eventually a maximum of $+2V$ from C_2, $-V$ from the transformer secondary, and $+V$ from C_1, i.e. C_3 will tend to charge up to $2V$ through diode DE. Allowance must be made for current drain from E to A, associated with the accelerated ions and inevitable insulation losses, and the maximum voltage is, therefore, a little less than V times the number of rectifiers.

The simple van de Graaff accelerator (van de Graaff 1931, Fig. 4.3(a)) consists basically of a moving loop of insulating material, at one end of which charge is sprayed on by a 'comb' of points maintained at some 30 kV. At the other end, inside a large hollow rounded conducting terminal, the charge is removed by a similar device. The removal process is an example of the Faraday-cage effect whereby a charged conductor inside a larger hollow conductor (the 'cage') will, when joined to the hollow conductor, transfer its charge whatever the potential of the cage. The positive ion source is located in this high-voltage terminal, and the ions are accelerated in an evacuated tube through a succession of disc-shaped electrodes with graded voltages supplied by a chain of resistors from the terminal down to earth potential.

In order to minimize the main voltage limitation, namely corona discharges to laboratory walls and nearby objects, the whole system is commonly enclosed in a pressurized steel tank containing a suitable gas, e.g. nitrogen with a few per cent of freon (CCl_2F_2), or sulphur hexafluoride (SF_6), at pressures up to 100 atmospheres. Axial voltage gradients of about $2\ MV\,m^{-1}$ are achieved. The radial voltage gradient, E_r, depends on the radius of the terminal r and the inner radius of the pressure vessel R,

$$E_r = V/(r\ln(R/r)),$$

with values of E_r possible up to about $20\ MV\,m^{-1}$. The condition for maximum V for a given E_r is easily shown to be $R/r = e$, for which $V = E_r r$. The spread of energy can be as small as 1 keV and the energy stability about $\frac{1}{2}$ keV.

In the tandem van de Graaff accelerator (Fig. 4.3(b)) a single terminal at high positive potential is used first to attract negative ions (e.g. H^-) from an ion source at earth potential and then, after a thin foil or gas has stripped them of two or more electrons, to repel the positive ions produced (e.g. protons) back to earth potential. Heavy positive ions can be multiply charged and the energies produced by a

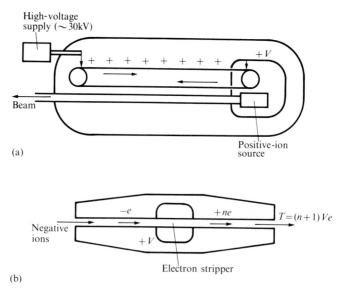

Fig. 4.3. (a) Schematic diagram of van de Graaff accelerator. V can be up to $\simeq 12$ MV. (b) Principle of tandem van de Graaff accelerator. V is maintained by a moving charged belt, as in (a).

terminal at a potential V can therefore be $(n+1)V$ eV if the negative ion is singly charged and the positive ion is a neutral atom less n electrons. The voltage is easily varied and can be very stable, and the current can be continuous or pulsed, making the machine ideal for coincidence counting or time-of-flight techniques in low-energy studies of nuclear structure and reactions.

The betatron (Kerst 1940) is essentially a transformer with a single-turn secondary that consists of the electrons under acceleration circulating in an evacuated toroid ('ring doughnut'). The increasing magnetic flux produced by the primary produces a tangential electric field and also constrains the electrons to move in a circle. The condition for simultaneous acceleration and an orbit of constant radius R is simply that the instantaneous magnetic flux density at the radius R should be always one-half of the mean flux density within the orbit. One bunch of electrons is accelerated in each period of a 50 or 60 Hz alternating current in the primary. A severe restriction on circular electron accelerators is the bremsstrahlung (radiation) loss (see p. 74) which is $(1/4\pi\varepsilon_0)$ $(4\pi e^2/3R)$ $(E/m_0 c^2)^4$ per turn, i.e. 8.85 $\times 10^4 E^4/R$ keV per turn for E in GeV and R in metres. This is clearly negligible for heavier particles.

The electron linear accelerator (Fig. 4.4) uses a cylindrical wave-guide fed with radio-frequency power in a mode that provides a series of regions of positive and negative electric field along the axis. These field regions move along the guide at a velocity designed by means of 'disc-loading' to be that of a bunch of electrons under continuous acceleration in a positive-field region. The 50 GeV Stanford electron linear accelerator is 3 km long, and is fed by 245 klystron amplifiers at a radio-requency of 2.850 GHz in pulses of length 2.5 µs. The

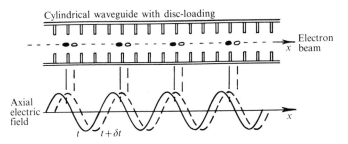

Fig. 4.4. Electron linear accelerator of travelling wave type. Bunches of electrons are shown solid at time t and hollow at $t+\delta t$.

repetition rate is 1–360 Hz, i.e. the maximum 'total pulse-on-time' is 1 ms in each second, a 'duty-cycle' of 0.1 per cent. The energy spread is about 1 per cent and currents up to 30 μA are possible.

An important development for linear accelerators is the use of superconducting waveguides that reduce considerably the r.f. power losses and, therefore, allow much higher duty cycles.

Repeated acceleration

The proton linear accelerator acts on a different principle from that of the electron linear accelerator. The earliest type (Wideroe 1928; Sloane and Lawrence 1931) used a series of 'drift' tubes of successively increased length along the accelerator axis, with alternate tubes connected to the terminals of a radio-frequency supply (e.g. 30 MHz). At a given time, the electric fields in alternate gaps were correctly directed for acceleration, and half a period later the bunches of particles were at the immediately next gap and, therefore, were again accelerated.

The modern version (Alvarez, 1946) uses a resonant cavity (Fig. 4.5) and the mode of oscillation of the injected radio-frequency power (~ 200 MHz) produces a unidirectional field along the whole length of the cavity, excepting where the drift tubes shield the axis. At a given time the fields at all the gaps between the drift tubes are favourable for

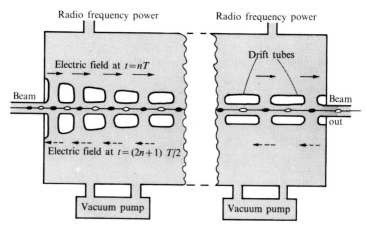

Fig. 4.5. Proton (or heavy ion) linear accelerator of resonant cavity type. T is the period of the applied radio-frequency power. Bunches of particles are shown solid for the times of acceleration ($t = nT$) and hollow when they are shielded by the drift tubes.

the acceleration of bunches of protons in the gaps. At unfavourable phases of the field the protons are within the metal drift tubes, and therefore shielded from the electric field. For $f = 200$ MHz, and $v = c/3$ ($\equiv 50$ MeV protons) the distance between gaps, and hence the length of a drift tube is 0.5 m. The power consumption is a problem, limiting the duty cycle to about one per cent, unless a superconducting resonant cavity is used.

A vital inherent property of the proton linear accelerator is its 'phase stability'. Consider the variation of the accelerating r.f. field with time, Fig. 4.6(a), for which there will be an optimum phase, point O, such that the energy increase per gap allows the particle to maintain the optimum phase throughout the machine. This optimum phase for a given type of particle, proton, or heavy ion, clearly depends on the distances between the gaps and the peak r.f. field. If, for any reason, a particle arrives late at a gap, point L_1, it will experience a greater than optimum field which will enable it to catch up a little on particles ahead of it in space at the optimum phase, hence at the next gap, one period later, it will have a phase L_2 which is relatively nearer to O. A similar consideration applies to early particles at E_1 which experience less than the optimum field and hence arrive relatively later, at E_2. Such behaviour after many cycles is often illustrated as in Fig. 4.6(b), which is an overlay of successive cycles to show the phases at successive gaps. Particles with phases near enough to the optimum will be trapped and their phases will oscillate about the optimum phase. This phase-stability, or axial-stability, is associated with radial instability or defocusing and it is usual to compensate for this by incorporating magnetic quadrupole focusing within the drift tubes (see p. 66). Since the particles are limited to a small range of phases near the optimum, it is evident that the beam current has a 'fine

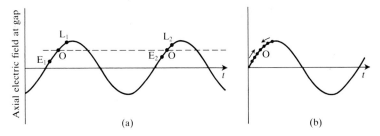

Fig. 4.6. Illustration of phase stability in resonant cavity proton (or heavy ion) linear accelerator.

structure' of 'bunches' of particles of duration about a fifth of the r.f. period, and for a frequency of 200 MHz this implies 1 ns bunches spaced at 5 ns intervals. The r.f. power duty cycle produces 'pulses' of duration about 200 μs at intervals of about 50 ms. Such fine and course time structures within the beam current are inconvenient for some experiments, e.g. coincidence-counting, but can be valuable for time-of-flight techniques. The particle energy for the linear accelerator cannot easily be varied and the total length becomes excessive for proton energies much above 100 MeV.

Proton linear accelerators usually have a Cockcroft–Walton injector and are themselves used as injectors for some synchrotrons (Fig. 4.9).

Orbital accelerators utilize the economy of space provided by the basically circular orbit arising from an applied magnetic field. The simple cyclotron (Lawrence and Livingston 1931, Fig. 4.7) relies on the fact that the period of revolution τ is independent of the velocity v of the particle. This follows from the familiar relation $qvB = Mv^2/R$. The period of revolution is $\tau = 2\pi R/v$ and therefore $\tau = 2\pi M/Bq$. Thus radio-frequency power supplied to the two D-shaped electrodes,

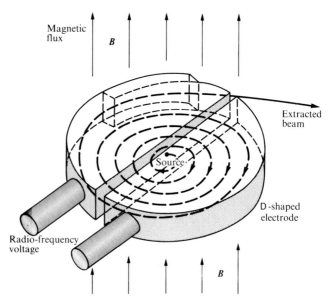

Fig. 4.7. Schematic diagram of simple cyclotron. The magnetic flux B is provided by a large electromagnet.

known as dees, will produce at time $t = 0$ an accelerating field for a particle of mass M charge q at a given gap, and when $t = \tau/2$ it will produce acceleration at the diametrically opposite gap provided the applied frequency f is equal to the 'cyclotron frequency' $1/\tau = Bq/2\pi M$. Similarly, there will be acceleration on its arrival at all subsequent gaps. When the electric field is unfavourable the particle is electrically shielded by the metal dees, but the magnetic field acts all the time. As the particle gains energy the orbit radius R increases until finally the particle is extracted at R_{max}, where the kinetic energy $T = q^2 R_{max}^2 B^2/2M$.

There is, however, a snag in the process at the higher energies since the particle mass is velocity-dependent, $M = M_0(1 - v^2/c^2)^{-\frac{1}{2}}$, owing to relativistic effects. Thus τ increases with v and, for a machine with a large enough radius, the particle would arrive at the gaps successively later than the times for optimum electric field until eventually the electric field at the time of arrival would be very small or zero.

At first sight the solution is obvious—increase B with R to keep in step with the increase of M with R. Unfortunately, this would lead to forces with an axial component that would push the particles away from the median plane (Fig. 4.8(a)). In fact, in the simple cyclotron B is made to *decrease* slightly with R in order to produce median-plane (or axial) focusing (Fig. 4.8(b)). The axial component of the force F produces axial focusing and the radial component constrains the particle to move in a (semi-)circle.

One way of overcoming the relativistic problem, which is used in the synchrocyclotron (McMillan 1945; Veksler 1945), is to concentrate on one bunch of particles at a time and gradually decrease the frequency of the applied radio-frequency power to allow the particles to arrive 'late' and yet still find an accelerating field. This turns out to be easier than it sounds because of an inherent phase stability whereby precise (and rapid!) orbit-by-orbit tracking of frequency and energy is not necessary. When one bunch of particles has been taken to the largest radius the frequency is increased to its original value and another bunch is accelerated. The radio-frequency varies from about 30 MHz to 15 MHz at a modulation frequency of 50 or 60 Hz, and hence there are 50 or 60 pulses of current per sec, each with an r.f. fine structure of bunches. The upper energy limit is set by the sheer size of the magnet which must produce a magnetic flux at all radii.

Phase stability in the synchrocyclotron requires the use of a different quarter-cycle of the r.f. compared with that for the resonant

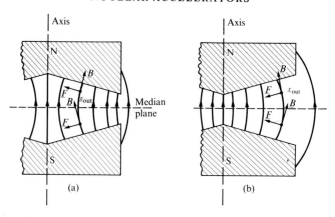

Fig. 4.8. (a) Defocusing effect of cyclotron magnetic field when B increases with radius. (b) Focusing effect when B decreases with radius. F is perpendicular to B and v. (Schematic and exaggerated.)

cavity linear accelerator. The period of rotation, $\tau = 2\pi M / Bq$, clearly increases with the mass M and, therefore, with the (linear) velocity or energy of the particle. In other words, the angular velocity decreases as the linear velocity increases, provided the radius R is free to vary. It follows that a particle that is 'early' in relation to the main bunch must be given more acceleration than the main bunch in order to make it arrive at the next gap relatively later. This conclusion, and similar reasoning for 'late' particles, means that phase stability is achieved in the second quarter-cycle of the sine wave for which the electric field between the dees decreases with increasing time.

The synchrotron (Oliphant 1943, Fig. 4.9) solves the problem of increasing magnet size encountered in the synchrocyclotron by using a fixed radius orbit and allowing the magnetic field acting on the orbit to increase as the particle momentum increases. Thus a ring magnet or series of magnets is used with a number of radio-frequency (r.f.) accelerating cavities arranged round the ring. Since the radius is essentially constant the frequency needs to increase if protons are to be accelerated from an injection energy of (say) 50 MeV, at which energy $v \simeq c/3$, but for electrons at a few MeV injection energy the velocity is already very close to c. Thus a proton synchrotron normally needs to vary both B and f in unison, but for an electron synchrotron (or a proton synchrotron with an injector of several GeV) only B needs to be varied and f is constant. The limit is now

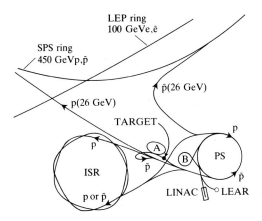

Fig. 4.9. Layout of accelerators at CERN, Geneva. The linear accelerator (LINAC) injects 50 MeV protons into the proton synchrotron (PS) which produces 26(-30) GeV protons (p). The proton synchrotron booster (B) is used to accelerate protons to 800 MeV and provides more efficient injection to the PS. Antiprotons (p̄) are produced by 26 GeV protons on the TARGET and those with energy 3.5 GeV are captured in the antiproton accumulator (A). Subsequently they are injected into the PS for acceleration to 26 GeV and hence to the SPS ring, where the counter-rotating p and p̄ are accelerated simultaneously to 450 GeV and focused for collisions in the centre of massive and elaborate detectors. The SPS ring has a diameter of 2.2 km. The large electron positron collider (LEP) ring has a diameter of 9 km. The low-energy antiproton ring (LEAR) receives 600 MeV antiprotons from the accumulator (A) and allows a variety of experiments in the range 100–2000 MeV. The intersecting storage rings (ISR) were the first to be used in p–p̄ collider studies.

purely financial for protons and heavier particles but for electrons there is the added problem of bremsstrahlung losses.

For conventional electromagnets covering 75 per cent of the orbit, with $B = 1.5\ T$, a proton energy of 450 GeV requires a diameter of 2.2 km (Fig. 4.9). The development of superconducting magnets with B five or six times larger has enabled the same radius machine to produce proportionately higher energies.

One of the important advances in the design of accelerators and beam transport systems was the application of the principle of strong alternating-gradient focusing (Christofilos 1952; Courant, Livingston, and Snyder 1952). The simple axial focusing of Fig. 4.8(b) is achieved by using a variation of B with R of the form

$$B/B_0 = (R_0/R)^n,$$

with the field index $n \simeq \frac{1}{2}$ or less. This type of focusing, which is conveniently found to be associated with simultaneous focusing in the radial direction, is weak and it led to relatively large pole gaps in the early proton synchrotrons to accommodate diffuse beams; consequently it required very large magnets. If n is increased beyond unity, in practice to about 300, the effect is much stronger, and axial focusing is now found to be associated with radial defocusing. However, a focusing lens (convex, f_1 positive) in conventional optics, combined with a defocusing lens (concave, f_2 negative) distance d away, has a net focal length F given by $1/F = 1/f_1 + 1/f_2 - d/f_1 f_2$. Hence, for $f_1 = |f_2| = f$, we have $F = f^2/d$. It follows that a combination of a sector magnet with positive n (B decreasing with R) and one with negative n (B increasing with R) produces focusing in both the axial and radial directions.

Many pairs of such magnets are used in so-called alternating-gradient machines and the focusing is so strong for the low energy particles at injection that they are constrained to a fixed path rather than one that allows different radii for 'early' and 'late' particles. In consequence, the phase-stability condition is at first the same as for a linear accelerator, i.e. in the first quarter-cycle of the sine wave; but later on the higher energies lead to different radii for early and late particles as in the synchrocyclotron and a transition needs to be made to the second quarter-cycle.

The use of alternating-gradient focusing in the proton synchrotron led to a considerable economy in magnet size, because of the much smaller beam diameters (millimetres rather than centimetres). It is also applied in the sector-focusing cyclotron which has B increasing with R to compensate for the relativistic effect and, therefore, can use a fixed-frequency r.f. supply. The defocusing effect of the increase of B with R is more than compensated by the alternations of magnetic-field gradient produced by radial hills and valleys on the pole pieces, hill facing hill and valley facing valley. In the limit the valleys become gaps between sector magnets with open spiral boundaries. Such machines yield essentially continuous currents of particles with quite high energies (up to approximately 600 MeV protons).

The transport of charged particle beams from accelerators to experimental equipment almost always involves the use of the focusing provided by a pair of quadrupole electromagnets (Fig. 4.10). Magnetic quadrupoles are also used in the drift tubes of linear accelerators, as supplementary focusing in synchrotron rings, and for the focusing of particle beams onto slits that form part of a momentum analyser based on the deflection and dispersion produced by a magnetic dipole, i.e. a magnetic spectrometer. Ion sources use quadrupoles (or sextupoles) to separate out the different magnetic dipoles of neutral atoms arising from the different orientations of the nuclear spin, thereby producing relatively high percentage polarizations for the injected and final accelerated beams.

From Fig. 4.10 it is clear that focusing is produced in the vertical plane and defocusing in the horizontal plane, and an axial particle is undeflected. If such a quadrupole of length L is followed at a distance d by a second quadrupole of identical form, but with the exciting current and hence the magnetic polarities reversed, then the overall effect of this quadrupole pair will be focusing in both directions, horizontal and vertical. The pole pieces are shaped to produce field components that are given by

$$B_x = by, \quad B_y = bx, \quad B_z = 0,$$

for which the deflecting force will be proportional to the distance from the $(z-)$ axis of the magnet. The focal length f for one such magnetic element is then approximately

$$f = Mv/qbL,$$

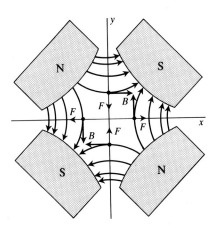

Fig. 4.10. Cross-section of a quadrupole magnet. The forces F are produced by the magnetic induction B for positively charged particles with a velocity in the z-direction, or at very small angles to z, and directed into the diagram.

where Mv is the particle momentum, and q its charge, and the overall focal length for the quadrupole pair is given, as before, by $F = f^2/d$.

Colliding beam techniques

The use of colliding beam techniques (Fig. 4.9) has produced a remarkable leap upward in effective energy. By 'effective energy' we mean the energy available in the centre-of-mass (CM) of the particle system. This is an important concept in classical mechanics, but becomes of even greater significance in collisions involving relativistic velocities.

Consider a particle, mass M_1 and kinetic energy T_1, colliding head-on with a particle mass M_2 initially at rest, i.e. $T_2 = 0$. Conservation of linear momentum and energy in the non-relativistic case leads to a kinetic energy for the centre of mass of the resulting system given by $T_{\text{of CM}} = T_1 M_1/(M_1 + M_2)$, and consequently the energy available within the system (i.e. in the centre-of-mass system) is $T_{\text{in CM}} = T_1 - T_{\text{of CM}} = T_1 M_2/(M_1 + M_2)$. For macroscopic colliding particles this energy in the centre-of-mass system could lead, for example, to a rise of temperature, emission of sound, or break-up. In nuclear collisions it can lead to excitation or break-up reactions. At sufficiently high energies new phenomena appear, notably the production of new particles such as pions (see pp. 125 ff), but only if the energy in the centre-of-mass is greater than $M_0 c^2$, where M_0 is the rest mass of the new particles.

For most particle-production calculations it is necessary to use relativistic mechanics. In the simplest case of collisions between identical particles $M_1 = M_2$, and with M_2 at rest the result is $T_{\text{in CM}} = \{(T_1 + 2M_1 c^2)2M_1 c^2\}^{1/2} - 2M_1 c^2$. In the high-energy limit $T_{\text{in CM}} \to (2T_1 M_1 c^2)^{1/2}$, and in the non-relativistic limit $T_{\text{in CM}} \to T_1/2$. Thus for proton–proton collisions with $T_1 = 30$ GeV and $T_2 = 0$, $T_{\text{in CM}} = 5.6$ GeV only.

For $T_1 = 450$ GeV, $T_{\text{in CM}} = 29$ GeV. These surprisingly low values of 'effective energy' are best appreciated by realizing that a highly relativistic proton colliding with a proton at rest corresponds to a heavy particle colliding with a light particle.

For head-on collisions between particles of the same rest mass and same kinetic energy, T, there is zero momentum of the centre-of-mass, and all the original kinetic energy is available in the centre-of-mass system, i.e. $T_{\text{in CM}} = 2T$. Thus, two 30 GeV protons colliding head-on yield $T_{in\ CM} = 60\ GeV$. But in order for this to be available in a collision with a proton at rest the energy of the incident proton would need to be

$$T_1 = (T_{in\ CM} + 2M_1c^2)^2/2M_1c^2 - 2M_1c^2,$$

i.e. $\qquad T_1 = 2038\ GeV = 2.038\ TeV.$

The colliding beam experiments with 30 GeV protons first undertaken at CERN, Geneva, used 'intersecting storage rings' (Fig. 4.9). Protons were 'stacked' in orbits in two highly evacuated rings intersecting at eight cross-over points. A switching system brought the counter-rotating beams into collision.

The collision of particles of opposite electric charge, for example electrons and positrons, and protons and antiprotons, is not only of special interest in testing theories for particle–antiparticle collisions but allows several simplifying techniques (Rubbia, McIntyre, and Cline 1976). Thus the same magnetic field that constrains electrons to circulate in one direction constrains positrons to circulate in the opposite direction, thereby facilitating collisions, and also allowing simultaneous acceleration. This was used for the SPS, Super Proton Synchrotron, at CERN (Fig. 4.9) with up to 450 GeV beams of protons and antiprotons in a 2.2-km-diameter ring. Storage of beams for up to 24 hours was achieved and collisions were in a volume about 30 cm long and 1 mm^2 in cross-sectional area. The arrangement is also the basis of LEP, the Large Electron Positron collider at CERN with, finally, 100 GeV beams in a ring of 9 km diameter (Fig. 4.9). Circulating beams of e and ē have their transverse oscillations about the ideal orbit damped by bremsstrahlung and contract to very small transverse dimensions. The TRISTAN eē collider in Japan has ~ 30 GeV beams. Conversion of LEP to a pp̄ collider with the present conventional 2 tesla magnets replaced by 10 tesla superconducting magnets would produce beams of energy between 8 and 9 TeV.

A linear accelerator for electrons can simultaneously accelerate positrons at regions midway between the electron bunches, since there the electric field is reversed, as required. This is the basis of SLC, the Stanford Linear Collider in the USA, with 50 GeV beams, which at the exit from the linear accelerator are deflected in opposite directions by the same magnet and then, on separate tracks, are brought into a 'one-shot' head-on collision.

For two 900 GeV proton/antiproton beams, which are available from the superconducting synchrotron at Fermilab (USA) $T_{\text{in CM}}$ $= 1.8$ TeV, and for one proton at rest the other would need to have $T_1 = 1.7 \cdot 10^{15}$ eV $= 1700$ TeV to produce the same value of $T_{\text{in CM}}$. A proposal in the USA to build a Superconducting Super Collider (SSC) with beams of 20 TeV protons would require a ring of 27 km diameter, would cost some 4.4 billion US dollars, and would take nine years to build. To obtain 40 TeV in centre-of-mass using a fixed target would require an incident beam of energy $8.5 \cdot 10^{17}$ eV. Protons in cosmic rays have been observed with energies estimated to be as high as 10^{24} eV, but they are extremely rare, and of course unpredictable, occurrences.

Problems

4.1. Calculate the maximum energy of electrons accelerated in a betatron with a maximum magnetic flux density of 1.0 T at an orbit radius of 1.0 m.

4.2. The proton beam entering a linear accelerator has an energy of 300 keV. Calculate the distance between the first and second accelerating gaps for an applied radio-frequency field of 200 MHz, given that the first 50 gaps accelerate the beam to 10 MeV. Identify necessary assumptions and any approximations.

4.3. Calculate the radial separation of adjacent orbits in a cyclotron for protons of energy 17 MeV at a radius of 0.6 m, assuming the effective voltage difference between the dees is 50 kV.

4.4. Calculate the ratio of the final to the initial radio-frequency for a 600 MeV proton synchrocyclotron, given that the necessary focusing is produced by a magnetic flux density at the final orbit which is 10 per cent less than the value at the centre.

4.5. (a) A 50 MeV beam of protons is to be deflected through 90° by a uniform magnetic field $B = 1.0$ T acting perpendicular to the beam over a circular area. Calculate the minimum diameter of the magnet. (b) The same magnet is to be used to deflect the beam through 45°. Calculate the minimum value of B required. (c) Show that the exit beam has a direction along a radius of the magnet whenever the entry beam is initially directed towards the centre of the magnet.

4.6. Derive the formulae given on p. 68 for the energy available in centre-of-mass of a particle of mass M_1, and laboratory kinetic energy T_1, incident on a stationary particle of mass M_2, (a) in the non-relativistic limit and (b) in the full relativistic form for $M_1 = M_2$. (Hint: For (b) use the invariance of $E^2 - p^2c^2$, where p is the momentum, which is zero in the centre-of-mass (CM) system, and E is the total energy.)

4.7. The energy of a beam of protons from a tandem accelerator is to be measured by a time-of-flight technique. Show that the error δE in the measurement of the energy E in the non-relativistic case is related to the error δt in the time of flight over a distance l by the equation

$$\delta E = \pm (2Ec\delta t/l)(2E/M_0c^2)^{\frac{1}{2}}.$$

Insert reasonable values of δt and find the corresponding values of δE for $l = 150\,\text{m}$ and $E = 25\,\text{MeV}$.

4.8. Electrons are injected at an energy T_0 into a travelling wave linear accelerator of length L, and experience an electric field X V m^{-1} at the optimum phase. Show that the accelerator appears to the electrons to have a length L' given by

$$L' = (m_0c^2/X)\ln\left(\frac{m_0c^2 + T_0 + XL}{m_0c^2 + T_0}\right).$$

Plot the variation of L' with T_0 for $T_0 = 0$ to $10\,m_0c^2$ for the Stanford accelerator with $L = 3\,\text{km}$ and a final energy of 20 GeV. (Hint: use the Lorentz–Fitzgerald contraction, $\delta l' = \delta l/\gamma$, with $\gamma = (m_0c^2 + T)/m_0c^2$.)

4.9. Deduce a reasonable radius and corresponding maximum energy of an electron synchrotron that radiates by bremsstrahlung 1 MeV per turn.

5. Nuclear instruments and methods

Energy loss by nuclear particles

The design of instruments for the detection of nuclear and atomic particles requires an understanding of how the particles transfer energy to the detecting medium. The same understanding is needed in the design of shielding against the harmful effects, physical and biological, of exposure to radiation, and in the planning of medically beneficial exposures. Some applications of isotopes, e.g. the monitoring of the thickness of thin films during production, also depend on the characteristics of particle absorption.

Charged particles and photons interact with electrons in matter, and to a very much smaller extent with the nuclei. Neutrons interact with the nuclei by means of the strong nuclear interaction, but scarcely at all with the electrons. Neutrinos interact with the nuclei through the weak interaction only, and are very difficult to detect; the probability of interaction is so low that a neutrino produced during beta-decay could traverse the diameter of the Earth 10^{11} times with a 50 per cent chance of survival.

Charged particles, other than electrons, lose energy at a rate shown in Fig. 5.1(a) and, except for the low-energy region, where multiply-charged particles (e.g. alphas) capture electrons, the following equation holds

$$-\mathrm{d}E/\mathrm{d}x = (q^2 n Z/\beta^2) f_1(\beta, \mathscr{I}),$$

where qe is the charge of the particle, n the number of atoms (each with Z electrons) per unit volume of the absorber, and $\beta = v/c$. The function f_1 of β and of $\mathscr{I}$, an appropriate mean of the excitation and ionization energies ($\mathscr{I} = 13.5Z$ eV), is essentially constant for kinetic energies less than $0.2 M_0 c^2$, where M_0 is the rest mass of the incident particle, for which region therefore $\mathrm{d}E/\mathrm{d}x$ is proportional to $1/\beta^2$. Notice that $\mathrm{d}E/\mathrm{d}x$ expressed as a function of β is, for a given absorber, independent of the mass M_0 of the charged particle. But the range R of a particle, obtained by integrating $\mathrm{d}E/\mathrm{d}x$, does depend on M_0, since

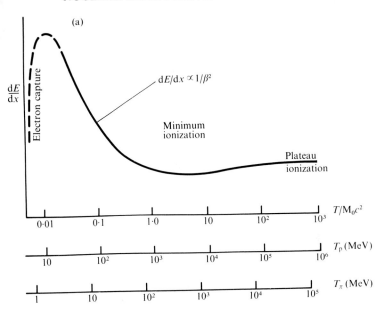

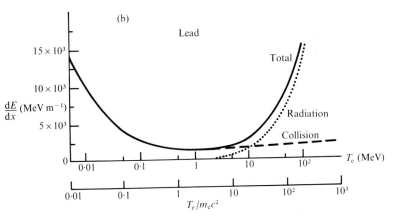

Fig. 5.1. (a) Variation of the rate of energy loss, dE/dx, with the kinetic energy (T) for 'heavy' charged particles (i.e. other than electrons) of rest mass M_0. Kinetic energy scales are shown for protons and pions. (b) Rates of energy loss for electrons in lead, showing the contributions from bremsstrahlung (Radiation) and ionization and excitation (Collision).

the appropriate variable is β and $dE = M_0c^2\beta d\beta$ (non-relativistic). For T between 10 MeV and $\sim 0.2\,M_0c^2$ we can thus assume dx is proportional to $(M_0\beta^3/q^2nZ)d\beta$, i.e. R is proportional to T^2/M_0q^2nZ. If we express R in kg m^{-2} (rather than metres), we need to multiply both sides of this proportionality by the density ρ of the absorber, but we note that $n = N_A\rho/A$, and very approximately A is proportional to Z, i.e. R (kg m^{-2}) is proportional to T^2/M_0q^2 and is approximately independent of the absorber material. A corresponding approximate empirical relation is $R = 0.034\,T^{1.8}/M_0q^2$ (kg m^{-2}), for T in MeV and M_0 expressed in proton masses. Notice the close parallel to Geiger's relation (see p. 21), expressed as $T_\alpha(\text{MeV}) = 2.12\,\{R(\text{cm of air})\}^{2/3}$.

Ionizing particles can impart considerable energy to struck electrons, and these secondary electrons, known as delta-rays, can themselves produce ion pairs and show their paths in suitable track detectors. When the primary particle is moving at high relativistic speeds, the number of delta-rays per unit length of primary track is simply proportional to q^2, and affords a means of particle identification. For large q values the end of the track has a diminishing dE/dx, because the particle begins to capture electrons and reduce its effective q (Fig. 5.1(a)), and this diminution can also be used to assess q. These measurements have been of considerable value in studies of the charge spectrum (and hence the mass spectrum) of high-energy primary cosmic rays, using nuclear emulsion detectors.

Energy loss by electrons involves not only excitation and ionization of atoms but also bremsstrahlung or 'braking radiation' (Fig. 5.1(b)). Whenever a charged particle is accelerated or decelerated it emits electromagnetic radiation. Multiple deflections of fast electrons by the Coulomb fields of nuclei in an absorber involve changes of momentum, and therefore involve accelerations. (It is the periodic accelerations of electrons in radio transmitting aerials that lead to radio-wave emission.) The rate of emission is given by the classical equation

$$-(dE/dt)_{\text{rad}} = (1/4\pi\varepsilon_0)(2e^2/3c^3)|d^2x/dt^2|^2.$$

For a particle of charge qe, mass M_0, approaching a nucleus of charge Ze, the force $M_0(d^2x/dt^2)$ is proportional to qZ/x^2, and hence $-(dE/dt)_{\text{rad}}$ is proportional to q^2Z^2/M_0^2. Clearly, the inverse dependence on M_0^2 makes the bremsstrahlung energy loss for protons quite negligible compared with its value for electrons. A complete analysis yields approximately $-(dE/dx)_{rad} \propto q^2 Z^2 T/M_0^2$ and for electrons this is the main term in the energy loss for $T > \sim 10$ MeV for heavy elements (e.g. lead, Fig. 5.1(b)) and for $T > \sim 100$ MeV for carbon.

Neutrons lose energy predominantly by elastic collisions with nuclei. At non-relativistic energies it is easy to show that the maximum energy transferred for a head-on collision of a neutron, initial energy T_0, with a nucleus of mass number A is

$$\Delta T = 4 A T_0 / (A + 1)^2,$$

hence materials of small A are more effective than those of large A. This is important in the choice of 'moderators' for nuclear reactors (see p. 112). Inelastic collisions also take place, and neutron absorption can be very important for $T_n < \simeq 1$ MeV.

Energy loss by gammas is possible by three main processes (1) photoelectric effect, (2) Compton effect, and (3) pair production. In the first the photon disappears, and its energy $h\nu$ is used to free an electron and give it kinetic energy. In the Compton effect the photon is scattered by an (almost) unbound electron, mass m_0, and the change in wavelength and hence the energy loss is related to the scattering angle θ by $\lambda_1 - \lambda_0 = (h/m_0 c)(1 - \cos\theta)$. The electron and positron (see pp. 140 ff) from pair production share the kinetic energy $h\nu - 2m_0 c^2$, and when later the positron annihilates with an electron at rest two photons are most commonly produced each of energy $h\nu_1 = m_0 c^2 = 0.51$ MeV. In all three processes energetic electrons are emitted and lose energy to the absorber by excitation, ionization, and possibly bremsstrahlung—the last being, of course, photons of (much) lower energy than the primary gamma-photon.

The relative importance of these processes as a function of photon energy, and how they differ for aluminium and lead, can be seen in Fig. 5.2. The intensity I_x of the radiation transmitted through a thickness x is given by $I_x = I_0 \exp(-\mu x)$ where I_0 is the incident intensity and μ is the linear absorption coefficient. It is important to note that I_x is the intensity of radiation with the same photon energy $h\nu$ as that of I_0, and there is always a 'build-up factor' to be considered in designing shielding and computing medical dosages to allow for the build-up of degraded photons of lower energy from the Compton effect and positron annihilation.

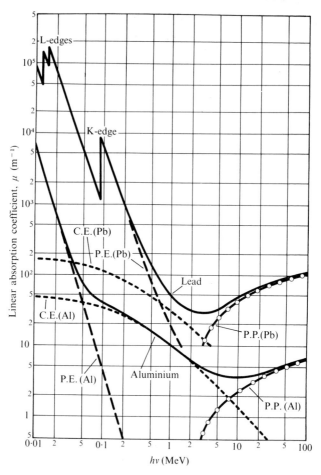

Fig. 5.2. Linear absorption coefficients for X-rays and γ-rays in aluminium and lead. The total coefficient μ is the sum of the three contributions; photoelectric effect (P.E.), Compton effect (C.E.) and pair-production (P.P.). Notice that both scales are logarithmic.

Radiation units and recommended dose limits

The total energy absorbed by a medium exposed to ionizing radiation is termed the *Absorbed Dose* and the S.I. unit is the gray or Gy ($1 \text{ J kg}^{-1} = 6.25 \, 10^{12} \text{ MeV kg}^{-1}$). The frequently used older unit the rad (from Radiation Absorbed Dose) is given by $1 \text{ rad} = 10^{-2}$ gray. In living organisms damage to tissues arises mainly from molecular disruption in individual cells in the track of the particle. The same total dose for particles with characteristically different types of ionization tracks produces different biological effects and a *Quality factor, Q,* (or Relative Biological Effectiveness, RBE) is introduced to allow for this in the definition of a '*Dose Equivalent*', DE. The biological effects also depend on the dose rate, and other factors, and to allow for this a further *factor, N*, has been introduced by the International Commission on Radiological Protection (ICRP) in its Publication 26 (HMSO, 1977). Dose Equivalent, measured in sieverts or Sv, is defined by

$$DE(\text{in Sv}) = \text{absorbed dose (in Gy)} \times Q \times N.$$

The earlier unit of dose equivalent, the rem (from Roentgen Equivalent Man—see later) is still frequently used and is defined by

$$DE(\text{in rem}) = \text{absorbed dose (in rad)} \times Q.$$

It follows that $1 \text{ rem} = 10^{-2}$ sievert for N set at unity. The quality factor Q (or RBE) is 1 for gammas and electrons, 2.3 for thermal neutrons, 5–10 for fast neutrons and 20 for alphas. The factor N has values in the range 1–10.

The earliest unit for radiation '*Exposure*' was proposed by Villard in 1908 and came to be known as the roentgen. It is not very convenient as it refers only to the ionization produced when X-ray or gamma photons are absorbed in air. An exposure of 1 roentgen (1 R) produces $2.58 \ 10^{-4} \text{ C kg}^{-1}$ of ions, and since an average of 32.5 eV is required to produce a pair of ions in air, 1 roentgen deposits $0.838 \ 10^{-2} \text{ J kg}^{-1}$, and is approximately equivalent to a dose of one rad $(= 10^{-2} \text{ J kg}^{-1} = 10^{-2} \text{ Gy})$.

For workers frequently exposed to nuclear and X-radiation, the maximum permissible dose, or 'tolerance dose' pre-1958, for whole-body radiation, has been successively reduced from 70 rem per year in 1925, to 50 rem y^{-1} in 1934, to 15 rem y^{-1} in 1950, and to 5 rem y^{-1} in 1956. The concept of 'tolerance dose', which implied a general threshold, and also the phrase 'maximum permissible dose', are now officially abandoned. The latest recommendation in the use of nuclear radiation was stated by ICRP in 1977, namely, all exposures to be kept *As Low As Reasonably Achievable* (ALARA). The aims are (a) to limit '*stochastic*' effects, i.e. those for which the probability of occurrence depends on the dose, e.g. cancers, and genetic effects, and (b) to prevent '*non-stochastic*' effects, i.e. those for which a threshold is believed to exist, e.g. cataract of the lens of the eye. The *dose equivalent limit* for radiation workers, assuming whole-body irradiation, is recommended to be 50 mSv y^{-1} (5 rem y^{-1}) for stochastic effects, with higher limits for certain organs. Based on non-stochastic effects, the dose equivalent limit for any organ is set at 500 mSv y^{-1} (50 rem y^{-1}), except the lens of the eye (150 mSv y^{-1}). Trainees under 18 years of age are set limits that are three-tenths of those for radiation workers.

An individual member of the general public is 'allowed' only one-tenth of the limits relating to radiation workers, e.g. a whole-body maximum for stochastic effects of 5 mSv y^{-1} (0.5 rem y^{-1}). For non-stochastic effects with uniform whole-body irradiation, the limit is set at 50 mSv y^{-1} (5 rem y^{-1}), and 15 mSv y^{-1} for the lens of the eye. There are detailed recommendations for other non-uniform irradiations, with distinctions between men and women. It is considered that the consequent average for the whole population should not exceed 1 mSv y^{-1} (0.1 rem y^{-1}). Radiation workers who might, in the course of their work, exceed three-tenths of the dose equivalent limits are *Classified persons* and are subjected to individual monitoring and mandatory medical surveillance. *Non-classified persons* who are unlikely to exceed three-tenths of the limit are not required to be given a pre-operational medical examination but are usually subject to personnel monitoring, and sometimes to medical surveillance.

The defined dose equivalent limits are over and above the radiation levels arising from natural background and medical applications of radiation, in spite of the fact that levels of background radiation in granite areas (e.g. Aberdeen) can be some five times the levels encountered in most of the rest of the country (e.g. in London, where in brick houses the gamma levels are about 0.4 mSv y^{-1}). The 1984 UK Review (HMSO 1984) gave revised values of doses due to natural radiation (all in mSv y^{-1}); cosmic rays 0.30, terrestrial sources 1.20 of which radon and thoron 0.80, potassium-40 and other isotopes in the body 0.37, with a total dose from nature of 1.87 mSv y^{-1} (187 mrem-y^{-1}). Medical irradiation leads to an average dose of about 0.25, of which more than two-thirds is from diagnostic techniques. Other man-made radiation doses are, on average, relatively small; nuclear weapons fall-out 0.01, occupational exposure 0.009 (of which less than half is from nuclear power stations), radioactive waste 0.002, and miscellaneous 0.011. In the UK the total man-made dose is ~0.28 mSv y^{-1} and the overall total is ~2.15 mSv y^{-1}.

In summary, the background and medical levels amount to about 2 mSv y^{-1}, other man-made levels total 0.03 mSv y^{-1}, and the recommended limit (for stochastic effects—cancer and genetic effects) for a radiation worker is 50 mSv y^{-1}, and 5 mSv y^{-1} for other people. It is useful to note that a dose-rate of 7.5 μSv h^{-1} for a 40-h week and 50 weeks per year constitutes an annual dose of 15 mSv, i.e. three-tenths of the annual whole-body dose limit for a radiation worker. The UK National Radiological Protection Board recommended in November 1987 that the limit for radiation workers should be reduced from 50 to 15 mSv y^{-1}, with comparable proportional reductions for other categories. At 15 mSv y^{-1} the risk of fatal cancer is now considered to be 1 in 2000 per year.

For a single whole-body dose of 4 Sv (400 rem) there is a 50 per cent chance of death within a short period; for 6 Sv (600 rem) an early death is almost certain. The effects of doses extending over many years are much less severe, and for both single and extended doses the somatic effects, i.e. those manifested in the body of the recipient, are more significant than the genetic effects that are revealed in succeeding generations.

Particle detectors

Most particle detectors utilize the electrons or positive ions produced in the detecting medium to generate an electric pulse. Minimum ionizing particles produce $\sim 10^5$ electrons per cm in liquids and solids and $\sim 10^2$ per cm in gases at S.T.P. In the *gas ionization counter* the electric field between two plates or grids collects all the primary ions, both positive and negative. The resultant pulse for one incident particle is very small, but the detector is more often used for integrating the pulses from a flux of X-rays or gamma-rays to produce a small, steady current.

The *gas proportional counter* uses higher voltages between the electrodes, and the primary ions gain sufficient energy to produce secondary ion pairs. Multiplications up to 10^6 are obtained for electric fields of a few hundred volts per centimetre, and since the output pulse is proportional to the initial number of ions useful information is obtained about the total energy of the incident particle, if it stops, or about its dE/dx and hence its identity if only part of its track intersects the sensitive region.

The *Geiger* (or *Geiger–Müller*) *counter* (see p. 20) uses even higher electric fields, and the release of ultraviolet photons by the numerous ions propagates the discharge throughout the counter volume, so that the size of the output pulse, which is large, is independent of the initial number of ions. It is a sensitive, robust, and useful instrument for simple monitoring of ionizing radiation, but by itself it gives no information on the identity or energy of the detected particle.

Pocket dosimeters (Fig. 5.3) are gas ionization counters that indicate the leakage of high voltage from an insulated electrode, rather like a gold-leaf electroscope. The total movement of a gold-plated flexible quartz fibre can be viewed against a scale by means of an eyepiece, and hence the integrated exposure estimated directly in microsieverts (or millirads or milliroentgens).

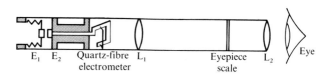

E_1 E_2 Quartz-fibre L_1 Eyepiece L_2 Eye
electrometer scale

Fig. 5.3. Schematic view of pocket dosimeter ($\simeq 12$ cm long). The lens L_1 forms an image of the quartz fibre on the eyepiece scale which is viewed through the lens L_2. The spring-loaded electrode E_1 allows recharging of the insulated electrometer by contact with E_2.

Solid ionization counters, consisting of a suitably treated semiconducting crystal (germanium or silicon), find increasing applications because solids are about 1000 times more dense than gases, and hence more compact, and furthermore they require much less energy to produce an ion pair ($\simeq 3$ eV compared with $\simeq 30$ eV for gases). The increased numbers of ions are of importance not so much for the larger pulses produced, since amplification is always needed, but because the mean percentage statistical variation ($100\, \delta N/N$ is proportional to $N^{-1/2}$) in the pulse size for a given energy deposited is significantly reduced, yielding excellent energy (or dE/dx) resolution. For low-energy particles, and for gamma-ray spectrometry, a germanium crystal (maximum volume $\simeq 100$ ml) with lithium diffused into it, to produce a carrier-free region, is an excellent detector except for the need to use and store it at liquid-nitrogen temperatures. An example of its resolution is shown in Fig. 5.4. Other types of *semiconductor detector*, the silicon surface-barrier and the n–p

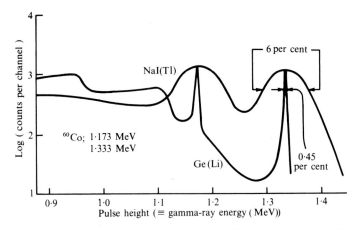

Fig. 5.4. Comparison of resolution achieved by NaI(Tl) scintillation counter (3″ × 3″ cyclinder) and Ge(Li) semiconductor detector for ^{60}Co gamma-rays. Note the two Compton scattering edges near 0.96 and 1.11 MeV in the Ge(Li) spectrum.

junction detectors, have relatively shallow ($\simeq 1$ mm) sensitive regions, but have the advantage that they operate at room temperature.

The present-day *scintillation counter* is a striking example of technological development of a simple device. In 1903, Crookes and, independently, Elster and Geitel discovered that the irradiation of zinc sulphide by alphas produced individual flashes of light rather than the expected steady luminescence. In this early work, a microscope and a dark-adapted eye were essential, but these have been replaced by an electron-multiplier, in which the photons are first converted to photoelectrons and then these electrons are repeatedly multiplied approximately fourfold by accelerating them through 100 V to a specially prepared 'dynode' that yields secondary electrons. The multiplication process is commonly continued by ten more stages ($4^{10} \simeq 10^6$, but up to 10^8 is possible) to yield useful electric pulses at the final collecting electrode or anode. The multiplication is roughly proportional to V^6, where V is the total voltage ($\simeq 1$ kV) across the dynodes, and therefore highly stabilized extra-high-tension (EHT) supplies are required for high-resolution analyses of pulse heights. To a good approximation the size of the pulse output is proportional to the light input, and this is proportional to the energy deposited by the nuclear particle or gamma photon in the scintillator. The most common scintillator for gammas is thallium-activated sodium iodide (NaI(Tl)) which has a rather slow release of the scintillation photons (time constant 0.25 µs). Plastic scintillators (and liquids) are available with time constants of a few nanoseconds, and are, used extensively in low- and high-energy research. The obtainable resolution is not so good as that of semiconductor detectors, but very large volumes of plastics with linear dimensions up to ≃ 1 m are available. Photomultipliers used in conjunction with these large scintillators introduce a time-jitter of only ∼ 1 ns and therefore allow fast timing and high resolution in coincidence counting.

Relatively large (≃ 2 litre) NaI(Tl) crystals, which are deliquescent and need to be encapsulated, have been produced for high gamma-detection efficiency. An example of the modest resolution for NaI(Tl) is shown in Fig. 5.4. Barium germanate crystals are also used.

Photomultipliers are essential in *Čerenkov detectors* which depend on the electromagnetic radiation, mostly visible or ultraviolet, produced by a charged particle when it enters a medium with a velocity that is greater than the (phase) velocity c/n of light in the medium, where n is the refractive index of the medium.

Part of the energy of the local pulse of electric polarization in the medium can be returned to the initiating particle by the coherent radiation produced by the relaxation of the polarization but only provided the particle velocity v is less than c/n, the velocity of that radiation. If $v > c/n$ the coherent radiation cannot catch up the particle and it escapes as Čerenkov radiation. This is the optical analogue of the sonic boom.

The intensity is small but the direction of emission of the light (Fig. 5.5) is at a characteristic angle θ to the track, given by $\cos\theta$

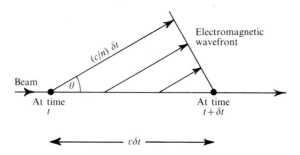

Fig. 5.5. Čerenkov effect. For particle velocity v greater than the phase velocity c/n, light is emitted parallel to the (generator of the) surface of a cone of half-angle θ.

$= (c/n)\delta t/v\delta t = 1/\beta n$. It is sometimes very useful to be able to make a direct measurement of θ and hence obtain the velocity of a particle.

The *ring Cerenkov detector* achieves this by a concave mirror focusing the light from the parallel cone surfaces (Fig. 5.5) to form a ring image with a diameter related to θ and hence to the velocity. The intensity of the ring, defined by dots of light produced by an image intensifier actuated by individual photons, depends on the nature of the charged particle and can thus allow simultaneous determinations of the identity, velocity, and direction of the particle.

Track detectors

There is no doubt that a picture (Figs 5.6 and 5.9) of the tracks of a nuclear event or 'explosion' carries more conviction for many people than complicated deductions based on pulse heights, pulse shapes, and counting rates from particle detectors. It is true that the time relationships, and even the sequence of events, are usually lost, but much information is recorded. Thus the density of the track can yield dE/dx and the delta-ray density can indicate charge; the range relates to the energy, mass, and charge, and the angles give linear momentum relations. If a strong magnetic field is used there is the added information (momentum/charge) provided by the curvature of the track. Although uncharged particles leave no tracks they can produce associated charged particle events and these, together with 'missing mass' and 'missing momentum' considerations, can lead to the identification of the uncharged particles.

The earliest track detector was the *Wilson cloud chamber* (1912), in which the positions of ions produced along the track of the charged particle in a gas are made visible by causing water droplets to form upon them (Fig. 8.2). This is achieved by allowing the particle to enter a chamber containing a gas saturated with water vapour and then suddenly expanding the gas by moving the piston-like floor (or ceiling or wall) of the chamber. The associated decrease of gas temperature produces supersaturation, and the ions act as favoured centres for the growth of drops. Stereoscopic flashlight photographs enable the three-dimensional characteristics of events to be measured. The diffusion-type cloud chamber has continuous sensitivity in a horizontal layer that is made supersaturated by an applied temperature gradient. Warm vapour diffuses downwards towards the base, which is usually cooled with solid carbon dioxide (dry ice).

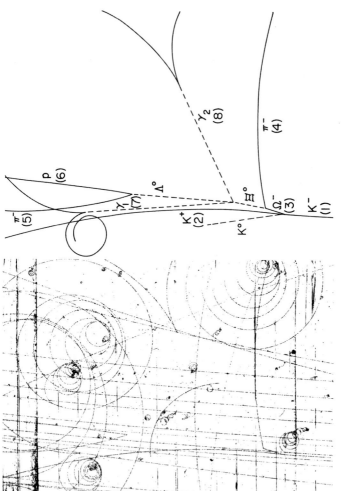

Fig. 5.6. Photograph from liquid hydrogen bubble chamber with a (perpendicular) magnetic field and the interpretation of the event of interest, which is the historic discovery of the omega-minus (see pp. 157 ff). The dashed lines are reconstructions of the trackless neutral–particles. The impressive spirals are from the electron-positron pairs produced by each γ-photon, and are explained by the steady loss of energy of the electron and the consequent steady reduction of track radius. (V. E. Barnes and 32 others, Phys. Rev. Lett. **12**, 204, 1964)

The *bubble chamber*, which utilizes the local heating along the track of a charged particle in a liquid, is particularly useful in conjunction with accelerators that produce pulses of particles at predetermined times. About 15 ms before the particles enter, a signal from the accelerator causes the pressure in the chamber to be suddenly reduced, in order to provide a superheated liquid. The local heating along the track produces bubbles which are photographed before the chamber is quickly recompressed to prevent boiling throughout the volume. Notice that it is not possible to 'trigger' the bubble chamber by using a signal (e.g. a scintillation) from the entering particle to actuate the pressure release. During the few milliseconds needed to effect the decompression the local heating would be dispersed and the track would disappear. Enormous bubble chambers have been built with dimensions of several metres (5 m at Batavia, USA), usually with liquid hydrogen, since the detecting medium is, of course, also the nuclear target for reactions and proton targets are of fundamental importance. Some chambers contain organic liquids or other higher-Z liquids in order to improve gamma-detection efficiency. The use of magnetic fields for determining momentum/charge, and also the sign of the charge, has been improved by the development of superconducting magnets built around the bubble chamber. The analysis of the stereo-photographs requires both skill and time, but the permanent record with very little inherent bias is of considerable value.

Special *nuclear photographic emulsions* had their heyday in the period 1945–65 in relation to research in cosmic rays and elementary particles. Large volumes ($\simeq 0.3$ m cube) of highly concentrated ultrafine-grain photographic emulsions were produced by stripping $\simeq 0.5$-mm-thick emulsions from their glass plate backing and stacking them with interleaving lens-tissue. For the tedious developing or processing they were first remounted on specially prepared glass to prevent undue distortion. Examination of nuclear emulsion tracks of about 1 μm resolution requires high-power ($\simeq \times 2000$) microscopes and large teams of observers, although some automatic scanning devices have been used.

Several other types of detectors have been based on the conversion of tracks of ions in a gas into a series of sparks between successive plate electrodes pulsed to sufficiently high potential differences. In the simplest form of '*spark chamber*' the spark tracks are photographed, but the delay and lack of precision in analysis are frustrating. In an improved version the acoustic pulse emitted by a spark is picked up by three or more microphones, and the times of arrival of the signals fed straight to a computer for 'on-line' analysis of the position of the spark.

The most extensive applications of 'instant' track detectors have been based on *wire chambers*. A planar array of very thin wires, covering areas of several square metres, is used to produce sparks between an adjacent plane of wires or a plane conducting sheet, or alternatively a pulse of current from the primary ions after a proportional-counter type of amplification. By means of multiple arrays of planes of wires with different orientations, resolutions of about 1 mm are possible. An alternative arrangement, known as a *drift chamber*, uses wires and planes of wires several centimetres apart. The initial position of the ions is pin-pointed by timing their 'drift' to the wires in the carefully designed uniform fields ($\sim 10^5$ V m^{-1}) before

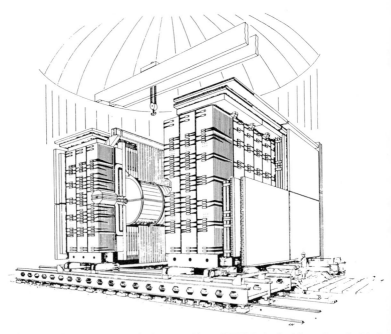

Fig. 5.7. The UA1 detector during assembly at CERN. In its final form the cylindrical central detector is totally enclosed by calorimeters which are in turn surrounded by double planes of muon drift chambers. A uniform magnetic field over the whole central track chamber is provided by aluminium coils.

entering the relatively high fields ($\sim 3 \cdot 10^7 \, V \, m^{-1}$) near the wires where the ion amplification takes place. Position along the wires is given by the division of charge to the two ends of the wire, but resolution in this direction is not at all so good. Cylindrical geometries are commonly used for colliding beam experiments, and by accurate positioning of up to 10^5 wires resolutions better than 0.1 mm have been achieved.

The current pulses in the wires are monitored by individual electronic devices that enable almost simultaneous analyses to be undertaken by a computer. These gas ionization detectors can be triggered either by a pulse from the accelerator that produces the primary-particle beam or by scintillation counters on each side of the array, provided that their signals are in time-coincidence. The use of such auxiliary detectors allows much more efficient experiments to be designed. Even so, one of the problems of nuclear research is the recording and analysis of vast amounts of experimental data.

Examples of the magnitude and complexity of detector arrays are given in Figs 5.7 and 5.8. They were designed for the UA1 and UA2

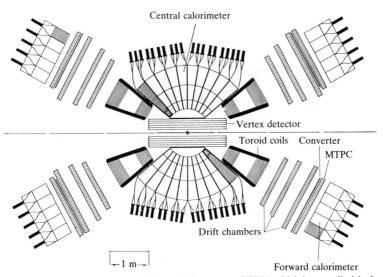

Fig. 5.8. Schematic plan view of the UA2 detector at CERN, which has a cylindrical symmetry about the beam direction. The central drift chamber is provided with a highly segmented calorimeter transverse to the beam. Other calorimeters relate more to the forward and backward directions and these detectors and associated drift chambers are provided with a limited region of magnetic field by toroid coils.

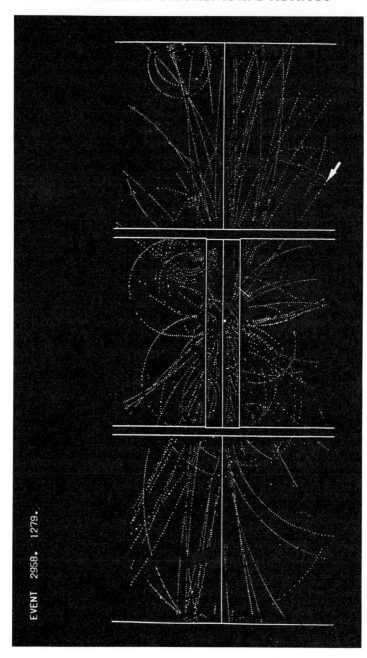

EVENT 2958. 1279.

Fig. 5.9. Charged particle tracks produced by a proton–antiproton collision at an energy of 540 GeV, as detected by the drift chambers of the UA1 experiment at CERN. The arrowed track indicates an electron from the decay of one of the first recorded examples of W-particle production.

(Underground area 1 and 2) experiments which led to the discovery of the W^+, W^-, and Z^0 particles (see pp. 170 ff). Features of both of these assemblies, each of over one thousand tons, are the 'calorimeters' or total absorption detectors, which are designed to measure energy. At particle energies above 10 GeV their energy resolution of about 1 per cent is better than that provided by magnetic analysis, and it improves with energy as $1/E^{1/2}$. Such a calorimeter consists typically of alternating layers of iron or lead, to produce secondary charged particles, and layers of plastic scintillator to produce the signal. Analysis of the scintillator pulses at successive layers can yield particle identification. An example of the excellent resolution provided by the drift chambers of the UA1 detector assembly is shown in Fig. 5.9.

Isotopes

Distinct nuclear species are commonly referred to as 'isotopes', whether radioactive or stable. Strictly speaking, isotopes are of the same element, i.e. in the same place (Greek: *topos*) in the classification table of the elements, and therefore have the same number of protons. Similarly 'isotones' are nuclei with the same number of neutrons, and 'isobars' are nuclei with the same atomic mass number A. Distinct nuclear species are properly termed 'nuclides' (sometimes nucleides), but we shall mostly use the popular term isotopes. Nuclear 'isomers' are excited nuclei with lifetimes greater than about 1 us. ^{137}Ba (2.6 min) and ^{110}Ag (250 days) are examples of the relatively few that have lifetimes longer than 1 s. An isomeric transition is the emission of a gamma from an isomer or isomeric state.

Isotopes were initially recognized in the three natural radioactive series and later identified by means of mass spectrometers, first by Thomson, confirmed by Aston (1920), for the neon-20, 22 isotopes. There are about 280 'stable' isotopes found in nature, including a few such as ^{40}K (0.0118 per cent of natural potassium) with a half-life of 1.3×10^9 years (β^-, e-capture), ^{204}Pb (1.48 per cent, 1.4×10^{17} years, α) and ^{238}U (99.27 per cent, 4.51×10^9 years, α). Within the naturally occurring radioactive series there are some 50 unstable isotopes, and as a result of nuclear-physics research many hundreds of artificial radioactive isotopes have been discovered, ranging from tritium (^{3}H, 12.26 years, β^-, 18 keV) right up to the new transuranic elements with atomic numbers 93–109 (see p. 108, and Table 6.1). Some of these, and some of the rarer stable isotopes, have found many applications in medicine and industry, as well as in chemistry, geology, and biology research. As a result, methods of preparation and separation have been developed in specialized supply laboratories.

The major sources of radioactive isotopes are (1) waste products from nuclear reactors, (2) irradiation of stable isotopes with neutrons within reactors, and (3) irradiation with particles from accelerators. Only longer-lived isotopes, e.g. ^{90}Sr(28 years) and ^{137}Cs(30 years), are usefully extracted from spent reactor fuel rods because of the tedious chemical separations involved. A large number of useful isotopes can be made by irradiation with neutrons in a reactor, and in nearly every case the radioactive product is of the same chemical element, e.g. ^{24}Na(15 hours, β^-) from ^{23}Na(100 per cent), and ^{60}Co(5.26 years, β^-) from ^{59}Co(100 per cent). This makes chemical separation of the radioactive isotope almost impossible, and therefore the specific activity (Bq kg^{-1} or Ci kg^{-1}) is often rather low. Szilard and Chalmers (1934) showed that if the irradiated element is in a chemical compound the gamma emitted after neutron capture causes a recoil that breaks the chemical bond. Free radioactive atoms are formed which readily allow chemical separation (e.g. in water) from the non-polar compound; e.g. ^{128}I(25 min, β^-) is formed from ^{127}I(100 per cent) in ethyl iodide. Deuterons are the commonest accelerated particles used in isotope production but the yields of isotopes are rather small, perhaps at most a few curies of activity. On the other hand, the product can be chemically dissimilar from the initial element and the easy separation can yield high specific activity, e.g. ^{22}Na (2.6 years, β^+, e capture), which is the longest-lived positron emitter, from ^{24}Mg (d, α)^{22}Na.

The separation of stable isotopes can be effected by a wide variety of methods, but all are expensive and inefficient. Magnetic separators give good mass resolution, but the limitations of their ion sources lead to very small yields, and the method is confined to the more exotic isotopes, e.g. new nuclides. Gaseous diffusion was the method selected, in the development of the nuclear bomb, for the separation of ^{235}U(0.72 per cent) from ^{238}U (99.28 per cent) in the form of the gas uranium hexafluoride (UF$_6$), there being only one stable isotope of fluorine (^{19}F). Strictly speaking, it was 'effusion' through the minute holes ($\simeq 10^{-8}$ m diameter) in thin sheets of a silver–zinc alloy, etched by hydrochloric acid. The rate is proportional to $1/M^{1/2}$, and some 4000 stages are necessary for 99 per cent ^{235}U. In thermal diffusion, using liquids or gases, lighter molecules move to regions of higher temperature, e.g. near a hot wire on the axis of a long (up to 35 m) vertical cylindrical tube. The method is useful for laboratory scale work. Centrifugation has been tried with promising results, and a recent development has been selective ionization of an atomic beam by tunable lasers, exploiting their narrow-wavelength bands and the marginally different energy levels of ^{235}U and ^{238}U, followed by electric deflection to separate the ionized from the un-ionized isotope.

The separation of deuterium (^2H, 0.015 per cent) from natural hydrogen is relatively easy and has great potential importance in relation to nuclear fusion reactors (see p. 113). The best large-scale process utilizes chemical exchange, a principle which can also be used for other low-A elements. Natural hydrogen gas and steam are passed up a vertical tube containing a catalyst. Cold water is passed down and by chemical exchange the deuterium from the natural hydrogen concentrates in this water. The enriched water is used to provide the gas and steam for the next stage. An alternative method is repeated electrolysis of water leading to concentration of deuterium in the residual water.

Applications of isotopes are many and widespread. In medicine their uses include both diagnosis and therapy. Thus a measure of the rate and location of the highly preferential uptake of $^{131}I(\beta^-, \gamma,$ 8.1 days) by the thyroid and of the rate of its excretion in the urine enables abnormal thyroid action to be studied. In cases of excess thyroid hormones in the blood (hyperthyroidism), a dose of ^{131}I ($\simeq 200\,\mathrm{MBq} \simeq 5\,\mathrm{mCi}$) yields relief in some 75 per cent of cases by virtue of the damage to the thyroid cells produced by the betas from the locally absorbed isotope. Some forms of cancer are treated by localized application of isotopes (e.g. ^{137}Cs ($\beta^-, \gamma,$ 27 years) in the uterus), and the development of powerful collimated ^{60}Co ($\beta^-, \gamma,$ 1.17 MeV, 1.33 MeV, 5.2 years) gamma sources ($\simeq 10^{17}\,\mathrm{Bq} \simeq 10^6\,\mathrm{Ci}$) provides for general radiation therapy the modern equivalent of the lower energy X-ray machines.

In industry, ^{60}Co gamma sources are used in the radiography of weldings and castings. Sterilization of food products, drugs, and medical instruments is sometimes undertaken, but massive doses ($10^4\,\mathrm{Gy} = 10^6\,\mathrm{rad}$) are required. The thickness of plastic and metal sheets can be monitored, during production, and thus controlled through feedback systems, by count-rates of transmitted or backscattered ($\theta \simeq 180°$) beta-particles. Transmitted alphas and gammas are also used, depending on the thicknesses and materials involved.

Leaks and blockages in underground pipes (e.g. oil pipes) can be traced by the introduction of radioisotopes. Other applications overlap the use of 'tracers' in chemical and biological research, i.e. the labelling of certain molecules with a radioactive isotope which, in general, has the same chemical properties as its stable equivalent. Thus, the wear on a radioactively doped or irradiated piston (or cylinder) can be estimated from the radioactive build-up in the lubricating oil, and rates of diffusion in metals and plastics can be measured even though the amounts of material involved are exceptionally small. Uniquely, the self-diffusion of an element can be studied by following one of its radioactive isotopes into a substrate of its stable isotope(s).

In organic chemistry and biology, ^{14}C (β^-, 155 keV, 5568 years) has many applications as a tracer, e.g. in exchange and transfer reactions. Owing to its continuous and apparently steady production in the atmosphere by cosmic rays, living plants and animals have very closely the same fraction of their carbon as ^{14}C, about 1 in 8×10^{11} atoms, yielding 14 β^- disintegrations per gram of carbon per minute. At death the exchange with the atmosphere ceases and radioactive decay sets in. Subsequent measurements allow the elapsed time to be estimated and dating of woods, for example, can be made for ages up to 50 000 years. Oxygen and nitrogen have radioactive isotopes with half-lives of seconds or minutes, much too short for general applications. It is then sometimes convenient to use stable but rare isotopes of these elements and trace their involvement in processes by mass spectrographic analyses. Another method is to activate the tracer by neutron bombardment in a reactor, a variant of 'activation analysis' of trace elements in geological and other specimens which depends on careful gamma-spectrographic analysis using, for example, Ge(Li) solid-state detectors.

The range of applications of isotopes is already very great, but it is certain that the possibilities are far from exhausted.

Problems

5.1. (a) What is the energy of a proton that has half the range of a deuteron of energy T? (b) What is the relation between the range and kinetic energy for particles with the same initial dE/dx?

5.2. Derive an expression for the energy lost by a neutron of initial energy T_0 in scattering at an angle θ by a nucleus of mass A. (Hint: Solve first in the centre-of-mass system and then relate $\cos \theta$ and $\cos \varphi$, where φ is the scattering angle in the centre-of-mass.)

5.3. Calculate the maximum energy imparted to an electron in Compton scattering of photons of energy 1.173 MeV and 1.333 MeV (see Fig. 5.4).

5.4. Use Fig. 5.2 to estimate the relative thicknesses of lead and aluminium to produce a given attenuation for photons of energies 10 keV, 100 keV, 1 MeV, 10 MeV, 100 MeV.

5.5. A Geiger–Müller counter has associated electronic circuits that 'paralyse' the counter for known adjustable times τ after each registered count in order to allow the ionization in the counter to be returned to zero.

(a) Show that the count-rate N_c after correction for the paralysis time is given by $N_c = N_0/(1 - N_0\tau)$, where N_0 is the observed count-rate.

(b) N_0 is recorded for a range of values of $\tau = 300 - 1200$ μs and $N_c = 2000\,\text{s}^{-1}$. Present the results as a straight-line graph, and deduce what you can from the slope and intercepts.

5.6. A Geiger counter records a rate, R_0, of particle detection of $60\,\text{s}^{-1}$ for a paralysis (or dead-) time, τ, of 500 μs. How long does one need to count to ensure that the statistical error is less than the dead-time correction?

5.7. A 5 MeV alpha-particle is stopped in a gas ionization counter. Assume that 30 eV is required to produce an ion pair and that all ions are collected, what will be the change in voltage if the capacitance of the counter is 50 pF?

5.8. Calculate the rest mass of the particles that begin to emit Čerenkov radiation in a medium of refractive index 1.5 when their kinetic energy exceeds 37 MeV (see Table 8.1).

5.9. A solution containing 0.05 μCi of ^{24}Na was injected into a patient's bloodstream, and after 5 hours the activity in 10 cm^3 of the patient's blood was found to be 140 min^{-1}. Calculate the total blood volume of the patient. (^{24}Na $\tau_{\frac{1}{2}} = 15$ hours.)

5.10. A high-energy charged particle beam strikes a target and the products are detected at a point outside the primary beam 30 m away from the target. Charged particles of energy 250 MeV arrive 63 ns after the gamma-ray pulses. Identify these particles.

5.11. Radiocarbon-dating assumes that the ratio of ^{12}C to ^{14}C in the Earth's atmosphere has remained constant at (about) 8×10^{11} during the periods in question. A sample of ancient wood is reduced to pure carbon, 1 g of which has an activity of 3.0 pCi (0.111 Bq). Determine the age of the wood. ($^{14}C \ \tau_{\frac{1}{2}} = 5568$ years.)

5.12. A sphere of plutonium-239, of mass 0.1 kg, is suspended in an enclosure at a temperature of 288 K. Given that its half-life is 24 400 years, and that the total disintegration energy in alpha-emission is 5.24 MeV, calculate (a) its activity and (b) its equilibrium temperature, noting the errors introduced by any simplifying assumptions. (Density of $^{239}Pu = 19\,000 \ \mathrm{kg\,m^{-3}}$, Stefan–Boltzmann constant $= 5.67 \times 10^{-8} \ \mathrm{J\,s^{-1}\,m^{-2}\,K^{-4}}$.)

5.13. A foil of aluminium-27, 0.02 mm thick, is bombarded by a 0.5 mA beam of 1 MeV protons. A proton–alpha reaction takes place with a cross-section of 20 barns. How long must the foil be bombarded in order to produce two micrograms of the stable product element?

5.14. Show that the activity of a foil containing N nuclei, after irradiation for a time t by a flux φ of nuclear particles is given by

$$A = \varphi \sigma N (1 - \exp(-\lambda t)),$$

where σ is the cross-section for formation of the product of radioactive constant λ, assumed to decay to a stable end-product.

6. Nuclear reactions

Introduction

Nuclear reactions, in the widest sense of this term, include (a) elastic scattering, (2) inelastic scattering, and (3) processes in which there is a change of identity in one or more of the nuclear particles involved (Fig. 3.1). Sometimes the term nuclear reaction is restricted to category (3), and sometimes to categories (2) and (3). These distinctions are usually obvious in any given context but need more careful consideration when the phrase 'total cross-section' is used. For example, the total reaction cross-section given by the optical model (p. 42) includes 'compound (nucleus) elastic scattering' (p. 99).

Several examples of reactions have already been introduced, e.g. in Chapter 3, where the need to conserve mass-energy was emphasized, and in Chapter 4 where the importance of centre-of-mass energy was introduced and the need to allow for relativistic effects was illustrated. In this chapter we shall develop some of these principles and give examples of nuclear reactions that have contributed to a deeper understanding of nuclear forces and nuclear structure and have led to important applications, such as nuclear reactors for energy production.

Energetics

Consideration of mass-energy conservation yields information on the energetics of a reaction. If the sum of the rest masses of the final particles is less than the sum of the initial masses the reaction is said to be 'exothermic', i.e. literally 'gives out heat', a rather misleading phrase.

For the reaction $A(a, b)B$, i.e. particle a on target A producing particles b (light) and B (heavy), we can write in terms of *nuclear* masses and the Q-value of the reaction

$$(m_A + m_a)c^2 - (m_B + m_b)c^2 = \Delta mc^2 = Q,$$

and hence Q is positive for an exothermic reaction, and appears as increased kinetic energy of $(B + b)$ compared with that of $(A + a)$. For a negative Q the reaction is said to be 'endothermic'.

Masses of nuclei as such are very rarely measured—exceptions are the very lightest nuclei, e.g. protons and alphas. Usually we measure masses of ions which are readily converted to masses of atoms by adding on the masses of the requisite small number of electrons. The internationally adopted scale of atomic masses identifies the mass of the ^{12}C atom as 12 u (exactly), where the symbol u stands for 'unified mass unit', the unification referring to the different scales used previous to 1961 by physicists and by chemists. When nuclear reactions are expressed in equation form it is usual to use symbols for nuclei, and before atomic masses can be used it is important to add and balance the requisite number of electrons. (Allowance for the mass equivalence of the electron binding energies is rarely required because the small Z differences between the atoms involve very small differences in electron binding energies.)

Thus the nuclear equation

$$_{Z_1}^{A_1}A + _{Z_2}^{A_2}a = _{Z_3}^{A_3}B + _{Z_4}^{A_4}b + Q$$

must conserve electric charge, and hence $Z_1 + Z_2 = Z_3 + Z_4$, as well as the total atomic-mass number $A_1 + A_2 = A_3 + A_4$. The corresponding atomic equation when A, a, B and b are all nuclei is

$$\left.\begin{array}{c} A \\ +Z_1e^- \end{array}\right\} + \left.\begin{array}{c} a \\ +Z_2e^- \end{array}\right\} = \left.\begin{array}{c} B \\ +Z_3e^- \end{array}\right\} + \left.\begin{array}{c} b \\ +Z_4e^- \end{array}\right\} + Q$$

and clearly the number of atomic electrons balances. Hence $Q/c^2 = (M_A + M_a) - (M_B + M_b)$, where M is the *atomic* mass.

However, not all the particles in the nuclear reaction need be nuclei. They can be, for example, neutrons, beta-particles, or neutrinos, and the result is then not always so simple. For the naturally occurring beta-emitters the emitted electrons are always negatively charged, i.e. ordinary electrons. Some artificially produced radioisotopes, however, are found to emit positively charged electrons i.e. positrons. A few isotopes emit both β^-- and β^+-particles, e.g. $_{29}^{64}Cu$ (β^- 0.57 MeV, β^+ 0.66 MeV, $\tau_{\frac{1}{2}} = 12.7$ h), and the atomic mass equations are then

$$\left.\begin{array}{c} _{29}^{64}Cu \\ +29e^- \end{array}\right\} = \left.\begin{array}{c} _{-1}^{0}e \\ -e^- \end{array}\right\} + \bar{\nu}_e + \left.\begin{array}{c} _{30}^{64}Zn \\ +30e^- \end{array}\right\} + Q^- \left(\begin{array}{c} 39\% \text{ of} \\ \text{decays} \end{array}\right)$$

and

$$\left.\begin{matrix} {}^{64}_{29}\text{Cu} \\ +29e^- \end{matrix}\right\} = \left.\begin{matrix} {}^{0}_{+1}e \\ +e^- \end{matrix}\right\} + \nu_e + \left.\begin{matrix} {}^{64}_{28}\text{Ni} \\ +28e^- \end{matrix}\right\} + Q^+ \left(\begin{matrix} 19\% \text{ of} \\ \text{decays} \end{matrix}\right).$$

Notice that the balancing of the added 'atomic' electrons leads to the cancellation of electron masses for β^--emission

$$Q^-/c^2 = M({}^{64}\text{Cu}) - M({}^{64}\text{Zn}),$$

but for the positron emission

$$Q^+/c^2 = M({}^{64}\text{Cu}) - M({}^{64}\text{Ni}) - 2m_e.$$

The neutrinos do not enter into consideration as they have zero rest mass, or less than $\sim 40 \text{ eV}/c^2$ (pp. 175 ff.). (See pp. 142 to 144 for the difference between the electron neutrino, ν_e, and the anti-electron neutrino $\bar{\nu}_e$). The Q-values therefore are equal to the maximum beta-particle energies respectively, with a small correction due to nuclear recoil.

Copper-64 happens to exhibit another type of 'radioactivity' known as 'electron capture', in which the nucleus captures one of the atomic electrons, usually from the K-shell. The result is

$$\left.\begin{matrix} {}^{64}_{29}\text{Cu} \\ +29e^- \end{matrix}\right\} + \left.\begin{matrix} {}^{0}_{-1}e_K \\ -e^- \end{matrix}\right\} = \left.\begin{matrix} {}^{64}_{28}\text{Ni} \\ +28e^- \end{matrix}\right\} + \nu_e + Q_{ec} \left(\begin{matrix} 42\% \text{ of} \\ \text{decays} \end{matrix}\right),$$

and therefore

$$Q_{ec}/c^2 = M({}^{64}\text{Cu}) - M({}^{64}\text{Ni}).$$

The energy Q_{ec} provides the kinetic energy of the neutrino and of the recoil ${}^{64}\text{Ni}$ ion and also the energy of atomic excitation of the nickel ion, which is revealed by the emission of characteristic X-rays, associated with a K- or L-vacancy in nickel.

For negative-Q (or endothermic) reactions it is necessary to provide energy to make the reaction possible. For an endothermic reaction A(a, b)B with target A at rest, the centre-of-mass kinetic energy $T_{\text{in CM}}$ needs to be greater than Q. Since some energy is needed to provide kinetic energy of the centre-of-mass $T_{\text{of CM}}$ in order to conserve the linear momentum of the centre-of-mass, we have

$$T_{\text{LAB}} = T_{\text{in CM}} + T_{\text{of CM}},$$

and since

$$T_{\text{in CM}} = T_{\text{LAB}} M_A/(M_A + M_a),$$

it follows that the condition for the reaction to 'go' is

$$T_{LAB} > Q(M_A + M_a)/M_A.$$

Inelastic scattering

Energy spectra of inelastically scattered particles, for example from (p, p′) experiments, provide information on the energy levels of target nuclei. Allowance must be made for the recoil energy of the struck nucleus, and this clearly depends on the angle of scatter.

Angular distributions of inelastically scattered particles are anisotropic and exhibit diffraction maxima and minima. Analyses of these distributions yield information on the angular momentum quantum numbers of the initial and final states of the target nucleus, and also on the deformation parameter describing the departure from sphericity of the target nucleus. Changes in angular momentum can be provided by the linear momentum and impact parameter of the incident particle—viewed classically, and in the quantum-mechanical treatment a further factor arises from the need to satisfy any difference in 'parity' between the initial and final states, since parity is conserved for the system as a whole in strong nuclear interactions.

The parity of a wave function representing a given state of a system is said to be positive (or even) if the sign of the wavefunction is unchanged as a result of reversing all the coordinate axes, $x \rightarrow -x$, $y \rightarrow -y$, $z \rightarrow -z$. If the sign is changed under this transformation, the parity is negative (or odd). States of nuclei have eigenfunctions with angular dependence given by spherical harmonics which are themselves of even or odd parity according to whether the orbital angular quantum number l of the state is even or odd. The parity is given by $(-1)^l$ and hence a change of l which is odd will be associated with a change of parity. An even change of l leads to no change of parity. If the incident particle has half-integral spin this can contribute to a change of orbital angular momentum l of ± 1, whilst conserving total angular momentum, simply by flipping its spin from $+\frac{1}{2}$ to $-\frac{1}{2}$, or $-\frac{1}{2}$ to $+\frac{1}{2}$, even for an impact parameter of zero.

Simultaneous observations of gamma-ray spectra, and angular distributions of gamma-rays in time coincidence, also provide information on energy levels and their spins and parities.

Types of nuclear reactions

There are basically four types of reactions in which the final particles differ from the initial particles:

(i) there is an exchange of a proton and a neutron, e.g. (p, n), (n, p), (^{3}He, ^{3}H), (^{3}H, ^{3}He);

(ii) the residual nucleus has fewer nucleons (the generic name for protons and neutrons) than the target nucleus, e.g. (p, pn), (p, 2p), (p, 2n), (p, 3n), (p, d), (n, 2n);

(iii) the residual nucleus has more nucleons than the target nucleus, e.g. (d, p), (d, n), (α, n);

(iv) high-energy reactions that lead to the production of particles other than nucleons and nuclei, e.g. mesons and hyperons and heavy leptons (see Chapters 8, 9).

To these we can usefully add nuclear fission (p. 106) which deserves a separate category, and also heavy-ion reactions using incident beams of Z-values up to that of uranium ($Z = 92$) which lead to a wide variety of final products.

Reaction mechanisms

(a) Compound nucleus reactions

We have already introduced (pp. 50, 52) the process of compound nucleus formation (Bohr 1936) whereby an incident particle enters the target nucleus and shares its kinetic energy and binding energy with many or all of the other particles. The subsequent processes during the lifetime of the compound nucleus, 10^{-20} s $< \tau < 10^{-14}$ s, depend on the total excitation energy or 'temperature' of the compound nucleus and include:

(i) re-emission of the same species of incident particle with the kinetic energy characteristic of elastic scattering—i.e. compound nucleus elastic scattering, leaving the target nucleus in its ground state,

(ii) re-emission of the same species of incident particle with less than the energy characteristic of elastic scattering, leaving a target nucleus in an excited state—i.e. compound (nucleus) inelastic scattering,

(iii) emission of one or more gamma-rays leading to partial or total de-excitation of the compound nucleus, which may or may not be radioactive,

(iv) emission of one or more charged particles, or neutrons, or both, with previous, interim and subsequent emission of gamma-rays all possible.

(v) nuclear fission and heavy ion reactions which are special cases of (iv).

The energies of the emitted particles for a compound nucleus of sufficiently high temperature are those expected for an 'evaporation' process, producing an approximately Maxwellian distribution of energies. In particular, the energy distributions are not dependent on the process of formation (Bohr's independence hypothesis), indicating that the compound nucleus forgets the parameters of its formation, except for the total energy and angular momentum, and also parity.

The angular distributions for an 'ideal' compound nucleus of non-zero l-value are not completely isotropic or spherical since any induced orbital angular momentum of the compound nucleus must lead to extra emitted particles in the forward and backward directions. A simple classical explanation follows from the realization that, for a target nucleus with $l=0$, incident particles with non-zero impact parameters will produce angular momentum vectors perpendicular to the beam direction, but no incident particle can produce an angular momentum vector parallel to the beam. Rather as mud flies off a rotating wheel, particles will preferentially populate the forward and backward directions, but there will be a deficiency of particles flying off in the directions around 90° to the beam. This is particularly obvious for heavier ion reactions which involve high angular momentum states of the compound nucleus.

(b) The optical model and reactions

Elastic scattering of an incident particle following an interaction with a target nucleus treated as a whole can be calculated from the optical model (p. 42) which uses a potential well of a given shape, size, and depth in the Schrödinger wave-equation. The process is termed 'shape elastic scattering', or 'potential scattering'. By making the potential well complex rather than real we introduce absorption, hence the phrase 'cloudy crystal ball'. The total reaction cross-section given by the optical model corresponds to an averaging over compound nucleus resonances. In practice this is useful, because at high enough energies the resonances become broader and closer together, and also the detectors used have insufficient energy resolution. The optical model reaction cross-section covers the wide variety of possible reaction processes, encountered at incident energies above about 10 MeV, including compound elastic scattering. To explain these we need to consider not only compound nucleus reactions but also the so-called 'direct reactions'.

(c) Direct reactions

The formation of the compound nucleus can be represented in terms of an initial reaction between the incident particle and a single nucleon of the nucleus. This creates a second 'free' particle and leaves a 'hole' in the state of the nucleus, a two-particle/one-hole state, known as a 'doorway state'. It is a doorway both to further particle-hole excitations leading to the compound nucleus and also to direct reactions. In these direct reactions the particle or cluster of particles initially struck escapes from the nucleus with or without the incident particle, or the nucleus is excited (inelastic scattering) with the escape of the incident particle. Such single stage reactions produce more particles at forward angles than expected from a compound nucleus. The energy distributions of escaping particles also show more particles of higher energy than for compound nucleus processes, especially at small angles. Furthermore, such groups of high energy particles, associated with excitation of low-energy states in the residual nucleus, show diffraction type angular distributions (Fig. 6.1) characteristic of a direct reaction. Such direct reactions take place in a relatively short time, about 10^{-21} s or less, whereas the lifetime of the compound nucleus ranges typically from 10^{-20} to 10^{-14} s.

The (p, 2p) direct reaction provides remarkably clear evidence of the energy states of protons within a nucleus. Viewed simply as a single-step process, a 'knock-out' reaction, one would expect the two outgoing protons to be at about $90°$ to each other, apart from relativistic corrections, and this is found to be the case. Furthermore, the sum of the energies of the two protons in time coincidence, and correlated near $90°$ angular separation, is the incident energy less the binding (or separation) energy of the initially bound proton, and less the relatively small recoil energy of the residual nucleus. If the ejected particle is the least tightly bound, the residual nucleus can be left in its ground state, but excited states are also observed. It is also possible to

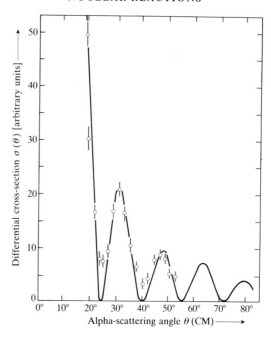

Fig. 6.1. Direct reaction inelastic scattering of 31.5 MeV α-particles by ^{24}Mg leading to the first excited state (2^+) at 1.368 MeV above the ground state. The theoretical curve is simply the square of the spherical Bessel function for $l = 2$, arbitrarily normalized. This 'plane-wave Born approximation' treats the incident particles as a plane-wave interacting with the nucleus at a characteristic radius, 6.4 fm, and assumes only a small perturbation (Born approximation) of the incident wave. (Watters 1956.)

eject nucleons from lower-energy states and the angular momentum of such nucleons, which is significant in the shell model of the nucleus (pp. 130 ff), can be deduced from a theoretical analysis of the angular distributions of the ejected particles. The probability of such knock-out reactions occurring is related to the probability of the target nucleus existing in a state resembling a free nucleon or cluster and a residual nucleus.

Nuclei favouring knock-out reactions also favour so-called 'pick-up' reactions of which (p, d) is the most common. In this case a neutron is picked up, with a probability that is enhanced when the momentum of the incident proton is close to that of the neutron in the target nucleus. Again the angular distributions relate to the angular momentum of the picked-up neutron and there are corresponding simple energy considerations.

Nucleons or clusters can be transferred both out of a nucleus (=pick-up) and into a nucleus. Introducing an extra nucleon or cluster by detaching it from an incident particle constitutes a 'stripping' reaction of which the most familiar examples are (d, p) and (d, n). This produces similar phenomena and information, and can be viewed as the inverse (or reverse) of the pick-up reaction. The probabilities of inverse reactions such as (d, p) and (p, d) are found to be identical, as expected by the statistical principle of detailed balance, provided the centre-of-mass energy is the same and provided allowance is made for the numbers of available final states which depend on the spins of the nuclei.

Heavy ion reactions

Ions of many nuclei with $Z \geqslant 3$ and $A \geqslant 5$, up to and including uranium ($Z = 92$), have been accelerated in heavy-ion linear accelerators, van de Graaff machines, cyclotrons, and even synchrotrons. They provide new areas of investigation, including the production of superheavy nuclei, new neutron-deficient and neutron-rich nuclei, excitation of states of very high angular momentum ($l \simeq 60$) which can fly apart under the centrifugal forces or produce cascades of gammas, Coulomb-induced fission, highly deformed nuclei, nuclear molecules, β-delayed proton emission, possibly two-proton radioactivity, and new reaction processes. A simple distinction between compound nucleus and direct reactions is not now tenable. Intermediate processes are needed to explain fast fission, and also the deep inelastic scattering that results from transfer of the incident kinetic energy and angular momentum to internal excitations of product nuclei of low kinetic energy.

In the range of energies 10–200 MeV per nucleon the relative ion–ion velocity passes through several characteristic nuclear velocities, namely the Fermi velocity (of a 'gas' of nucleons), the 'nuclear sound' velocity, and the pion-production threshold. Phenomena expected include nuclear shock-waves, phase transitions at a conjectured pion condensation within the nucleus, and a phase-transition from nuclear matter to a quark-gluon plasma. ^{16}O and ^{32}S have been accelerated in the CERN SPS synchrotron to allow nuclear collisions at energies up to 225 GeV per nucleon to be studied. At Brookhaven there is a well-developed proposal to collide ions with atomic masses up to 200 at energies up to 100 GeV per nucleon in each beam. The state of nuclear matter formed could be similar to that of the whole Universe a few microseconds after the initiation of the Big Bang (pp. 174 ff).

Meson, hyperon, and antiproton reactions

The production and identification of new particles in high-energy nuclear interactions are treated in Chapters 8 and 9 but here we consider studies of the interaction of beams of such particles with nuclei. Pion–nucleon interactions are of central importance in nuclear physics and elastic and charge-exchange scattering experiments, especially with polarized nucleon targets, provide fundamental evidence on nuclear forces. Pion–nucleus reactions probe the structure of nuclei, in particular the distributions of neutrons and protons in space, and can produce exotic nuclei.

Kaon interactions can produce hypernuclei in which a nucleon is converted into a hyperon, for example a lambda (Λ) or a sigma (Σ). With K^+ beams the interaction is weaker than with K^- and hence the nuclear interior can be more readily explored.

Antiproton beams are now available with several accelerators for a wide range of energies. They provide information on the nucleon–nucleon interaction, especially its short-range nature, and on proton–antiproton bound states in nuclear matter.

Neutron reactions

The zero electric charge of the neutron makes it a particularly useful particle for probing the nucleus since there is no Coulomb repulsion to overcome. However, for the self-same reason, it is relatively difficult to detect its presence and hence to measure its energy, and it is particularly difficult to provide neutrons in beams of good energy resolution and small angular spread. Fermi and his co-workers in 1934 noted that neutrons slowed down by elastic collisions with nuclei in hydrogenous materials are much more reactive than 'fast' neutrons produced by alpha-particles on beryllium (see p. 52). Most nuclei absorb slow neutrons to yield artificial radioactive isotopes, often producing gammas at the same time.

The simplest reaction can be expressed in the general form

$$^A_Z X + ^1_0 n \rightarrow ^{A+1}_Z X + Q_1,$$

where Q_1 is typically 4–10 MeV and the kinetic energy of the neutron ($\simeq 1/40$ eV) is in comparison small enough to be neglected. The excess energy appears as excitation of ^{A+1}X, and most commonly is emitted as gammas, hence the (n, γ) reaction is often termed 'radiative absorption'. But ^{A+1}X has excess neutrons compared with the initial ^{A}X and is usually β^--active,

$$^{A+1}_{Z}X \rightarrow {}^{A+1}_{Z+1}Y + {}^{0}_{-1}e + \bar{\nu}_e + Q_2.$$

An important example is $^{115}_{49}In$ (abundance 95.7 per cent), yielding ^{116}In which is β^--active with a half-life of 54.1 min, producing stable $^{116}_{50}Sn$. Indium foils are used to measure integrated fluxes of low-energy neutrons by means of the beta-activity developed.

Studies of neutron absorption cross-sections as a function of energy in the electron-volt range revealed rapid and large changes reminiscent of resonance phenomena. Fig. 6.2 shows the result for natural indium and illustrates the exceptionally large cross-sections ($> 10^4$ b) and the very small energy spreads or widths ($\Delta E < 1$ eV). These

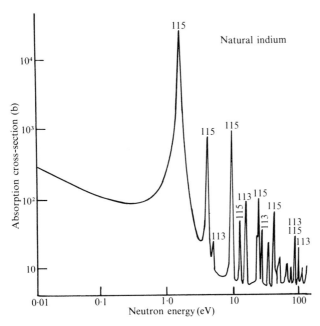

Fig. 6.2. Resonant absorption of low-energy neutrons by natural indium (95.72 per cent ^{115}In, 4.28 per cent ^{113}In).

neutron resonances arise when the mass of the target nucleus plus that of the neutron plus the mass equivalent of the neutron kinetic energy in centre-of-mass is very close to the mass of the compound nucleus in specific quantized energy states. Since the energy to separate a neutron from an unexcited nucleus is 4–10 MeV this energy must also be the excitation energy of the compound nucleus formed by neutron absorption. At such high-energy excitations the compound nucleus has energy states every few electron-volts, and very small adjustments of neutron energy are needed to pick out the levels. At higher neutron energies (> 100 eV) the peak widths become broader, and from Heisenberg's uncertainty relation $\Delta E . \Delta t \geqslant \hbar$ it follows that the mean lifetimes of the excited states must be shorter. The 'decay' of the excited state is either by gamma-emission or by particle-emission. For neutron re-emission, leaving the residual nucleus in its ground state, we have compound elastic (or resonance) scattering. For $\Delta E = 1$ eV, $\tau \simeq 6 \times 10^{-16}$ s—rather short for gamma-emission but remarkably long for neutron emission, since we expect about 10^{-20} s or less. This is because the excitation energy is shared between many or all of the nucleons in the compound nucleus until a statistical fluctuation makes it reside on one neutron.

Other important examples of neutron resonances are shown for ^{238}U in Fig. 6.3.

Fission and chain reactions

Fermi recognized that in radiative absorption of neutrons the net result, following the expected beta-emission, would be a nucleus with Z increased to $Z + 1$, and he proceeded in 1934 to investigate the element with the highest known Z, i.e. uranium ($Z = 92$), hoping to produce new 'transuranic' elements. The expected beta-activity was found to have several characteristic half-lives, and, since chemical tests suggested that the Z-values were not between 85 and 92, it was concluded that the new element 93 (and possibly 94) had been produced. Other workers made detailed chemical analyses which indicated how complex was the range of reaction products, and in late 1938 Hahn and Strassmann were forced to the conclusion that barium, lanthanum, and cerium, with nuclear charges 56, 57, and 58, were among the elements formed. Within a few days Meitner and Frisch (1939) had realized the significance of earlier suggestions (e.g. Noddack 1934, and Hahn and Strassmann 1939) and wrote 'it seems

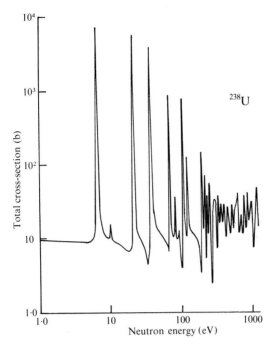

Fig. 6.3. Resonant absorption of low-energy neutrons by ^{238}U. This total cross-section includes contributions from both (n, n) and (n, γ) reactions. The five largest resonances are almost entirely (n, γ), elsewhere (n, n) is dominant.

possible that the uranium nucleus has only small stability of form, and may after neutron capture, divide itself into two nuclei of roughly equal size.' They pointed out that these fission fragments would have excess neutrons and be beta-active, and each fission process would release considerable energy, about 200 MeV, nearly ten times greater than the highest known energy in a non-fission reaction (22 MeV for ^{6}Li(d, α)^{4}He).

Within a few more days Frisch had detected high-energy fission fragments, and very soon a number of workers confirmed the startling discoveries. Fermi suggested that neutrons might be simultaneously emitted. It was found that so-called delayed neutrons were emitted for a short time after the initiating neutron bombardment had ceased and the intensities had characteristic half-lives up to about 1 min. The delay is associated with the half-life for beta-emission, which produces a nucleus that immediately ($< 10^{-20}$ s) emits a neutron. Later it was found that on average about 2.5 'prompt' neutrons are emitted, i.e. at the moment of fission. At an early stage it was expected that this number would be greater than unity, and the possibility of using these neutrons to produce further fissions and hence a chain reaction was considered in March 1939 by several research groups, including von Halban in France and Fermi in the USA.

It was realized that if, say, two new neutrons were effective for each initial neutron then after only 80 generations of multiplication there would be some 10^{24} effective neutrons. These could release about 5 $\times 10^6$ kilowatt-hours of energy (equivalent to about 5000 tons of TNT). From 10 kg of uranium, in a time estimated at less than 10^{-6} s, the result would be a tremendous explosion. Such speculations were extended to cover the possibility of controlled chain reactions, but after the outbreak of the Second World War (in late 1939) it was deemed necessary to restrict the publication of results.

After the War had been brought to an end by the two nuclear bombs exploded over Japan in 1945, the essential information for the production of nuclear bombs and nuclear-power generation was gradually released. It was confirmed that new elements $Z = 93$ (neptunium, Np) and $Z = 94$ (plutonium, Pu) were indeed produced in the U(n, γ) reaction (p. 105), as Fermi suspected. Using a variety of nuclear reactions all the possible 'transuranic' elements with atomic numbers from $Z = 93$ to 109 have now been produced, and claims exist for higher numbers. (See Table 6.1.) An 'island of stability' is expected for nuclides near $Z = 114$, $N = 184$.

The mass distribution of the uranium fission fragments (Fig. 6.4) revealed that the fission was predominantly asymmetric, a typical

TABLE 6.1

Transuranic elements (longest-lived isotopes)

Neptunium	$^{237}_{93}$Np $(2.14 \times 10^7$ y)	Nobelium	$^{259}_{102}$No (1 h)
Plutonium	$^{244}_{94}$Pu $(7.59 \times 10^7$ y)	Lawrencium	$^{260}_{103}$Lr (3 min)
Americium	$^{243}_{95}$Am $(7.95 \times 10^3$ y)	Kurchatovium	$^{261}_{104}$Ku (63 s)
Curium	$^{247}_{96}$Cm $(1.67 \times 10^7$ y)	Hahnium	$^{262}_{105}$Ha (40 s)
Berkelium	$^{247}_{97}$Bk $(1.4 \times 10^4$ y)		^{263}106 (0.9 s)
Californium	$^{251}_{98}$Cf $(7.9 \times 10^2$ y)		^{262}107 (0.12 s)
Einsteinium	$^{254}_{99}$Es $(4.8 \times 10^2$ d)		^{265}108 (1.8 ms)
Fermium	$^{257}_{100}$Fm (10.0 d)		^{266}109 (1 ms)
Mendelevium	$^{258}_{101}$Md (54 d)		

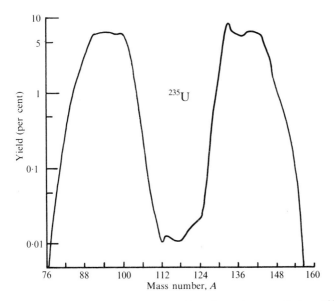

Fig. 6.4. Distribution of masses of fragments from thermal neutron fission of ^{235}U.

example being

$$^{235}_{92}U + {}^1_0n \rightarrow {}^{140}_{56}Ba + {}^{93}_{36}Kr + 3({}^1_0n).$$

12.8 days $\downarrow \beta^-$

$^{140}_{57}La$

40.2 hours $\downarrow \beta^-$

$^{140}_{58}Ce_{82}$ $\qquad$ $^{93}_{41}Nb_{52}$

Five β^- emissions ($\tau_{1/2} = $ 2 seconds to 10^6 years)

The fact that ^{140}Ce has 82 neutrons and ^{93}Nb has 52 neutrons is significant since nuclei with Z or N values near certain 'magic' numbers are known to be particularly stable—these magic numbers are 2, 8, 20, 28, 50, 82, and 126 neutrons or protons (see pp. 130 ff).

The energy released per fission can be estimated from the variation of nuclear binding energy with atomic-mass number. The total binding energy B (see p. 29) is defined by

$$B = \{Zm_p + (A - Z)m_n - m(A, Z)\}c^2,$$

and is therefore the energy required to break up the nucleus (A, Z) into its constituent isolated neutrons and protons. A plot of B/A, the binding energy per nucleon, is shown in Fig. 6.5, and if we consider B/A values of 7.5 MeV for $A = 235$, 8.3 for $A = 140$, and 8.5 for $A = 93$, then the energy released is $-B_{235} + B_{140} + B_{93} = -(235 \times 7.5) + (140 \times 8.3) + (93 \times 8.5) = 190$ MeV. Calculations for the above reaction using atomic masses give 200 MeV, and measurements for ^{235}U give an average value for all types of fission of (204 ± 7) MeV. About 165 MeV appears as kinetic energy of the fission fragments and the rest as the kinetic energy of neutrinos, gammas, betas, and neutrons.

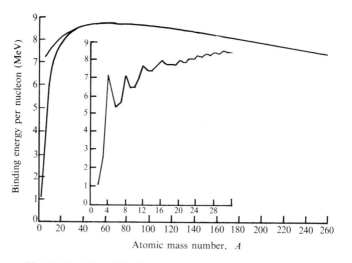

Fig. 6.5. Variation of binding energy per nucleon with mass number.

Natural uranium consists of 99.27 per cent ^{238}U, 0.72 per cent ^{235}U, and 0.0057 per cent ^{234}U. It is found that ^{235}U undergoes fission with low-energy 'thermal' neutrons ($T_n \simeq 1/40$ eV), but ^{238}U, the common isotope, requires neutron energies of at least 1 MeV (Fig. 6.6). The main reason is that a neutron that is paired off with another neutron is more tightly bound than an odd neutron in a comparable nucleus. Thus a thermal neutron entering ^{235}U forms a neutron pair and the total consequent excitation energy (which must be removed by gammas, say, if the resultant nucleus is to be left in its ground state) is 6.5 MeV, but for ^{238}U forming ^{239}U it is only 4.8 MeV. The

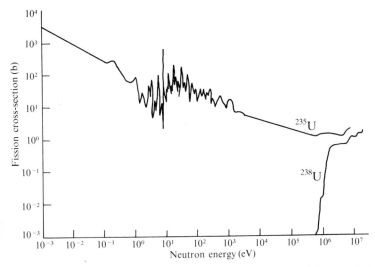

Fig. 6.6. Cross-sections for fission as a function of neutron energy for ^{235}U and ^{238}U.

separation of the fission fragments is something like alpha-emission from a heavy nucleus, although the mass (and charge) ratios are rather different and the potential barrier is consequently different. The wave-mechanical potential-barrier penetration probability is sensitively dependent on the available energy (p. 49; ^{238}U in Fig. 6.6).

This suggests that natural uranium is unsuitable for a nuclear bomb, since most of the neutrons emitted in fission have energies less than 1 MeV, but there is a further factor that makes it totally unacceptable, namely, the high (n, γ) cross-section of ^{238}U leading to $^{239}_{93}$Np and $^{239}_{94}$Pu (Fig. 6.3). For the first atomic bomb dropped in the Second World War pure ^{235}U was used, after laborious separation from ^{238}U (see p. 90); and for the second one, $^{239}_{94}$Pu. Plutonium is produced by ^{238}U (n, γ) and in turn decays by alpha-emission with a half-life of 24 360 years, yielding ^{235}U.

$$^{238}_{92}U + n \rightarrow {}^{239}_{92}U \underset{\beta^-}{\rightarrow} {}^{239}_{93}Np \underset{\beta^-}{\rightarrow} {}^{239}_{94}Pu \underset{\alpha}{\rightarrow} {}^{235}_{92}U$$

The odd neutron in ^{239}Pu ensures that it undergoes fission with thermal neutrons, and, therefore, with neutrons of all energies. Since plutonium is chemically different from uranium, it is relatively easy to separate.

The possibility of a controlled nuclear chain reaction using natural uranium was shown to depend on minimizing the loss of neutrons of energies 5–100 eV in ^{238}U (Fig. 6.3). The solution adopted, and still used in thermal reactors, was to form the natural uranium in relatively thin rods embedded in very pure graphite or heavy water (deuterium oxide). The elastic scattering by these light nuclei quickly slows down or 'moderates' the fast neutrons from the uranium rods, with very little absorption. Ordinary water cannot be used with natural uranium because of the relatively high neutron absorption cross-section for protons, leading to deuterons and 2.2 MeV gammas. The thermal neutrons diffuse back into the fuel rods where rather less than 50 per cent produce fission in ^{235}U. The multiplication factor k for neutron numbers per generation needs to be slightly greater than unity in order to build up a suitable neutron density, and hence power production, and the level can be controlled by the insertion or extraction of control rods of cadmium, boron or other elements that have a very high (n, γ) cross-section for thermal neutrons. If all neutrons were prompt the build-up would be uncontrollably rapid, since a generation has a period of only $\sim$ 1 ms, but, fortunately, about 0.4 per cent are delayed by beta-decays with half-lives greater than 0.1 s. Hence k is adjusted to be just less than unity for prompt neutrons and is able to be just greater than unity when all neutrons are included.

An important factor for both reactors and bombs is the critical size. Neutrons are lost from the surface of the device at a rate approximately proportional to the square of the linear dimensions, whereas neutron generation rates increase as the cube of the size. For natural-uranium thermal reactors the total size is of the order of a few metres cube; for bombs the uranium or plutonium needs to be of the order of 5 cm radius. Note that such a lump of ^{235}U would be spontaneously explosive, since only one initiating neutron is required and there are always a few neutrons present from spontaneous fission or from cosmic rays. In a bomb, therefore, the critical size is achieved by the use of initially well-separated subcritical masses with conventional explosives to effect rapid assembly of a single lump.

During the last thirty years several distinct types of nuclear reactor have been designed and operated for the generation of electrical power and for research purposes. Some use natural uranium enriched in ^{235}U or ^{239}Pu, which allows smaller reactor sizes and the use of ordinary water as a moderator. Some are specifically designed to produce more 'fissile' material (e.g. ^{233}U from ^{232}Th) than is consumed whilst still producing useful power. Other types of these thermal breeder or 'convertor' reactors, using ^{238}U, do not quite break even. The so-called 'fast' reactors use ^{239}Pu because of its relatively large fission cross-section for 'fast' neutrons and they can therefore dispense with the moderator. This type is more efficient in neutrons and hence allows more efficient breeding by conversion of ^{238}U to ^{239}Pu. For thermal-reactor power generation, the heat produced at temperatures up to 800 K is transferred from the fuel rods to a heat exchanger by circulating fluids (coolants), e.g. carbon dioxide or helium. Ordinary water for both moderating and cooling can be used in enriched reactors that are small enough to be transportable, or for use in ships and submarines. For fast reactors liquid sodium is probably the best heat-transfer fluid.

The relative safety of different types of nuclear power stations is an important and still unresolved consideration. The majority of accidents appear to be attributable to human error rather than to poor design or unexpected failures of materials. On the other hand, economic factors limit the number of safety devices that can be incorporated into a reactor. The generally accepted principle that benefits must outweigh the risks highlights the problems of assessing the risks—and also the benefits.

Fusion reactions

In nuclear fission a heavy nucleus of lower B/A is converted into (two) nuclei of higher B/A and considerable energy is released (Fig. 6.5). Similarly, when two light nuclei ($A < 10$) of lower B/A combine to form a heavier nucleus of higher B/A, energy is also released, and this reaction is known as nuclear fusion. For example, in the deuterium–tritium reaction:

$$^2_1\text{H} + ^3_1\text{H} \rightarrow ^4_2\text{He} + ^1_0\text{n} + 17.6 \text{ MeV}.$$

Because of Coulomb repulsion, deuterons need energies greater than 1 keV before the cross-section (i.e. probability) is large enough to make this reaction observable.

Such energies can be achieved in a high-temperature plasma, i.e. totally ionized gas, and it is believed that the energy produced could compensate for energy losses and be drawn off for power generation. Such thermonuclear reactors have been investigated for over 35 years, but many problems remain, notably the problem of how to contain the hot dense plasma for long enough intervals. The Lawson criterion states that the product of n, the number of ions per unit volume, and τ, the containment time, must be greater than $10^{20}\,\mathrm{m}^{-3}\,\mathrm{s}$ for the deuterium–tritium reaction in order that the useful power out should exceed the input power. Thermal neutrons at room temperature have a most probable energy of $kT = 1/40\,\mathrm{eV}$. The temperature corresponding to $1\,\mathrm{keV}$ is therefore about $12 \times 10^6\,\mathrm{K}$, and the difficulties are considerable. Most devices that have been tried are based on carefully designed magnetic fields in either linear or toroidal vessels. The fields cause the moving ions to stay away from the walls, at least during pulses lasting about $1\,\mathrm{ms}$. So far a controlled and adequately sustained reaction has not been achieved, although progress is promising as, for example, on the Joint European Torus (JET), (Rebut and Keen 1987).

An alternative approach to containment uses minute droplets, pellets or hollow spheres of deuterium and tritium illuminated by $10^{-9}\,\mathrm{s}$ pulses of high-power laser light. The surface layers ablate, i.e. are burned off, and the recoil of the lower layers compresses the pellet to very high densities at temperatures high enough to produce nuclear fusion. The time taken for the compression or implosion to become an explosion provides an inertial containment time τ. A major problem is to find an adequately efficient laser system of the optimum wavelength.

Nature provides high densities and high temperatures within stars, of which the Sun is a good average example. Although the surface temperature is only about $6000\,\mathrm{K}$ the Sun's internal temperature is up to $15 \times 10^6\,\mathrm{K}$, and densities of the predominantly hydrogen plasma are up to $130 \times 10^3\,\mathrm{kg\,m}^{-3}$. Under these conditions the reactions suggested by Bethe (1939) are almost certainly the dominant source of solar energy, the first having such a low cross-section that it has not yet been observed in the laboratory:

$$\mathrm{^1_1H} + \mathrm{^1_1H} \to (\mathrm{^2_2He}) \to \mathrm{^2_1H} + \beta^+ + \nu_e + 0.42\ \mathrm{MeV},$$

$$\mathrm{^2_1H} + \mathrm{^1_1H} \to \mathrm{^3_2He} + 5.49\ \mathrm{MeV},$$

$$\mathrm{^3_2He} + \mathrm{^3_2He} \to (\mathrm{^6_4Be}) \to \mathrm{^4_2He} + \mathrm{^1_1H} + \mathrm{^1_1H} + 12.86\ \mathrm{MeV}.$$

The net result is

$$4(\mathrm{^1_1H}) \to \mathrm{^4_2He} + 2\beta^+ + 2\nu_e + 26.7\ \mathrm{MeV}.$$

An alternative series of reactions involves carbon, nitrogen, and oxygen nuclei with the same net result, and for the Sun the two series are probably equally important.

An explosive thermonuclear or fusion device was developed in the USA, using a fission bomb to produce the high temperatures, and was first exploded in 1955. In principle, such hydrogen bombs can be made exceptionally powerful since the 'burning' is self-sustaining and the amount of fuel, which is inherently stable, is not restricted, as it is in the fission bomb, by considerations of critical size. They can also generate enormous quantities of radioactivity in the atmosphere, with consequent dangers to life on Earth.

Problems

6.1. A proton at rest absorbs a 'thermal' neutron (of negligible kinetic energy) to form a deuteron and a gamma photon. Calculate the energy of the photon and the recoil energy of the deuteron. ($M_H = 1.007825$ u, $M_n = 1.008665$ u, $M_D = 2.014102$ u, 1 u = 931.5 MeV/c^2.)

6.2. A nucleus of mass m_1, kinetic energy T_1, is incident on a nucleus of mass m_2 initially at rest. The compound nucleus formed breaks up into nuclei of masses m_3 and m_4 which are emitted along the line of the incident beam. Derive an expression for the Q of the reaction in terms of T_1 and T_3 and the ratios m_1/m_4 and m_3/m_4.

6.3. In a non-relativistic nuclear reaction A(a, b)B the particle b is ejected at 90° to the direction of the incident particle a, with a kinetic energy T_b. Show that the Q of the reaction is given by

$$Q = T_b(1 + M_b/M_B) - T_a(1 - M_a/M_A).$$

6.4. (a) Make a rough estimate of the lowest energy of alpha-particles that would be expected to exhibit diffraction effects in elastic scattering by ^{64}Ni (use non-relativistic formulae; mass of alpha-particle = 6.68×10^{-27} kg). (b) The threshold energy for the (α, n) reaction on ^{64}Ni is 5.20 MeV. What is the mass of the stable nuclide formed? The masses of the other atoms involved are 4.002603 u and 63.927958 u, and that of the neutron 1.008665 u. (c) The nuclide formed in the (α, p) reaction on ^{64}Ni is unstable. Without attempting numerical calculations, give reasons for the type of radioactivity that it can be expected to exhibit.

6.5. The nucleus $^{15}_{7}N$ has an excited state at 11.90 MeV. What is the energy of neutrons leading to a corresponding resonance in the reaction $^{14}_{7}N$ (n, p) $^{14}_{6}C$? (Atomic masses: $^{14}N = 14.003074$ u, $^{15}N = 15.000108$ u; neutron mass 1.008665 u.)

6.6. A nucleus of mass m captures a K-electron and emits a neutrino. Show that the nuclear recoil is $T_m = Q^2/2mc^2$ and that almost all the available kinetic energy is associated with the neutrino.

6.7. The isobars $^{37}_{18}A$ and $^{37}_{17}Cl$ have atomic masses 36.966772 u and 36.965898 u respectively. Does $^{37}_{18}A$ decay by electron capture or positron emission, and what is the Q-value of the reaction?

6.8. Given that only 0.1 mm of ^{113}Cd is needed to reduce a flux of thermal neutrons to 1/10000th of its original value, deduce the effective cross-section of ^{113}Cd for the absorption of thermal neutrons. (Density of cadmium $= 8.7 \times 10^3$ kg m^{-3}.)

6.9. Show that the mass of uranium needed to produce a megawatt-day (24 hours) of energy by nuclear fission is approximately 1 g. (Q for fission $\simeq 200$ MeV.)

6.10. Calculate the mass of deuterium needed to produce a megawatt-day (24 hours) of energy by (indirect) nuclear fusion to form alpha-particles. ($Q \simeq 25$ MeV.)

6.11. Calculate the Coulomb repulsion energy for fission fragments from $^{235}U + n$. Assume the fragments to be spherical, with their surfaces touching and of mass ratio 1.5. (Use the radius formula on p. 43.)

6.12. (a) Calculate the rate of change of the mass of the Sun given that its surface temperature is 5700 K and its radius 7×10^8 m. Note the necessary assumptions. (b) Assume that the mass loss is attributable to the (indirect) fusion of protons to produce alpha-particles. ($Q \simeq 25$ MeV.) What fraction of the Sun's proton content is changed to helium in (i) 10^6 years, (ii) 5×10^9 years? (Mass of Sun $= 2 \times 10^{30}$ kg.)

6.13. Show that the minimum gamma-ray energies required to produce (a) proton recoils of energy 5.7 MeV, and (b) nitrogen nucleus recoils of energy 1.4 MeV are relatively large and significantly different. Show that these recoils can be produced, however, by neutrons of approximately the same energy and that such neutrons are available from the reaction $^{9}_{4}\text{Be} + {}^{4}_{2}\text{He} = {}^{12}_{6}\text{C} + {}^{1}_{0}\text{n}$, with $T_{\alpha} = 5.3 \text{ MeV}$ ($M_{\text{Be}} = 9.012186 \text{ u}$, $M_{\text{He}} = 4.002603 \text{ u}$, $m_{\text{n}} = 1.008665 \text{ u}$, $1 \text{ u} = 931.5 \text{ MeV/c}^2$). (See p. 52.)

7. Nuclear forces and models

Introduction

The inverse-square law of force for gravity was recognized by Newton in the seventeenth century, and the similar relation for electrostatics was identified by Cavendish and by Coulomb about 100 years later. It is not uncommon when first approaching other forces, for example, Van der Waals forces and the nuclear force between a neutron and a proton, to hope for comparably simple power laws. Unfortunately, such an approach is rarely successful, and alternative methods must be adopted, many of which find expression in terms of the variation of the potential energy with the distance between the two particles. For the inverse-square laws the potentials vary quite simply as the inverse of the separation (Fig. 7.1(a)) and the application to the electronic structure of the atom, for example, is relatively straightforward.

The solutions of the Schrödinger wave equation for the ground state of the hydrogen atom are summarized in Fig. 7.1(b). Notice that the wavefunction φ is spherically symmetric in this case and can be written $U(r)/r$. The magnitude of $|\varphi|^2$ is often referred to as the probability density (or even the electron or charge density, but this can be misleading), and a careful distinction needs to be drawn between 'probability per unit volume' ρ_V (Fig. 7.1(c)) and 'probability per unit radial increment' ρ_R (Fig. 7.1(d)) as functions of radius. To emphasize this point we show in Fig. 7.2 the corresponding quantities for a sphere of uniform density, e.g. a billard ball. Notice that the ρ_R of Fig. 7.2(b) does not give the 'feeling' of a billiard ball. Nevertheless ρ_R is more common than ρ_V in representations of atoms and nuclei.

For nuclear forces no single potential has been found with the same generality as those for gravity and electrostatics, but advances in understanding have been made by assuming simple forms for the interaction potential. We shall use the simplest possible, the square well, to consider the earliest identified nuclear-force problem, the proton–neutron interaction, first as a bound system, i.e. the deuteron, and then unbound, as in low-energy neutron scattering by protons. Next we shall present Yukawa's important suggestions that led to the

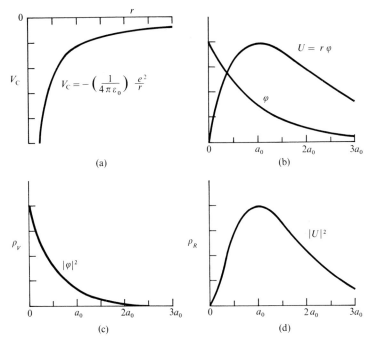

Fig. 7.1(a) Coulomb potential energy of the electron in a hydrogen atom; (b) wavefunction $\varphi = \varphi_0 \exp(-r/a_0)$ and radial wavefunction U, for the ground state of the hydrogen atom; (c) probability per unit volume; (d) probability per unit radial increment. (These diagrams are all drawn to scale, with a_0 the Bohr radius for $n=1$, $a_0 = h^2\varepsilon_0/\pi\mu e^2$ with $\mu = m_0(M_p/(M_p + m_0))$, the reduced mass of the electron. The maximum of U is at $r = a_0$ and equals $a_0\varphi_0/e$, with e the base of natural logarithms and $\varphi_0 = (1/\pi a_0^3)^{1/2}$.)

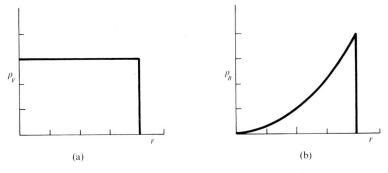

Fig. 7.2. Densities or 'probabilities' for a billiard ball: (a) volume density; (b) radial density.

discovery of mesons and to a theoretical basis for internucleon and other nuclear potentials. The so-called weak interaction encountered in beta-emission requires different approaches, and will be treated rather briefly. Finally, we note the special difficulties presented by real nuclei, because of the many-body problem, and discuss some of the models used to interpret nuclear properties and nuclear reactions.

The deuteron

In some ways the proton–neutron combination in the deuteron is the nuclear equivalent of the proton–electron in the hydrogen atom. In the atom it is a good approximation to refer to the proton as the centre of the system, because it is 1836 times heavier than the electron. But the proton and neutron have the same mass (within about 0.2 per cent), and we need to refer the potential and the wavefunction to the coordinates (r, θ, φ) in the centre-of-mass, where r is the internucleon distance, and use the reduced mass $\mu \simeq m_p/2 \simeq m_n/2$ in the wave equation,

$$\frac{d^2 U(r)}{dr^2} + \frac{2\mu}{\hbar^2} \{E - V(r)\} U(r) = 0.$$

Notice that we have already assumed for simplicity that the solution is spherically symmetric with $\psi = U(r)/r$, and we now need to examine $V(r)$ and E. For $V(r)$ we shall assume the simple spherically symmetric square-well potential (Fig. 7.3(a)) and hope to find something about its depth V_0 and range b. The available relevant information about the deuteron is its binding energy $B_d = 2.225$ MeV and its root-mean-square (charge) radius, $R_{rms} = 2.1 \pm 0.06$ fm, from high-energy electron-scattering experiments.

    The energy in the ground state is clearly $E = -B_d$ (Fig. 7.3(a)). This single piece of information cannot provide preferred values for both V_0 and b, but only a relation between them. From the solution of the wave equation, given in Appendix C, the relationship is as shown in Fig. 7.4 in the curve labelled $B_d = 2.225$ MeV.

    Using the second piece of information, R_{rms}, a corresponding relation between V_0 and b can be found (Appendix C). The relation is also shown in Fig. 7.4 for $R_{rms} = 2.10$ fm, together with the relations for R_{rms} set at the limits provided by the experimental error ± 0.06 fm. The intersection of the B_d curve and that for $R_{rms} = 2.10$ fm yields the values $V_0 = 34.6$ MeV and $b = 2.07$ fm. Notice that $b = 2.07$ fm is close

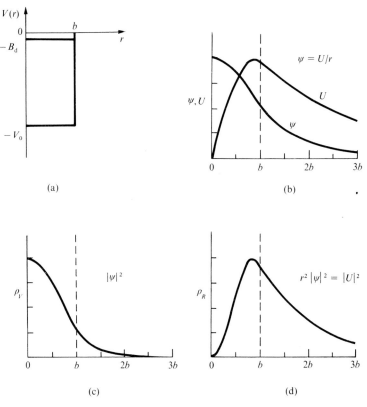

Fig. 7.3(a) Simple square-well nuclear potential for the deuteron as a function of the relative position coordinate r, with the bound-state level of energy $-B_d$ (approximately to scale); (b) wavefunction ψ and radial wavefunction U for the deuteron, scaled to the same maximum value; (c) probability per unit volume for the deuteron; (d) probability per unit radial increment for the deuteron.

to the value of $R_{rms} = 2.10$ fm, but this is entirely fortuitous as shown in Appendix C. Notice also that the values of V_0 and b are rather sensitive to the errors in R_{rms}, with the very small error in B_d (± 0.00020 MeV) quite unimportant, and all we can really deduce from these two data are

$$V_0 = 34.6 \pm^{10.3}_{6.5} \text{ MeV} \quad \text{and} \quad b = 2.07 \pm 0.30 \text{ fm}.$$

The forms of U and ψ, and $|U|^2$ and $|\psi|^2$ are shown in Fig. 7.3 for $V_0 = 34.6$ MeV and $b = 2.07$ fm. In general, they are not markedly

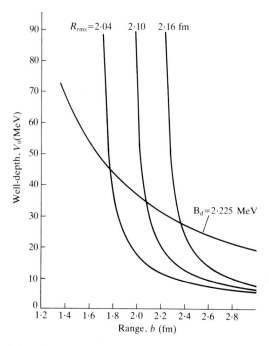

Fig. 7.4. Variation of the deuteron square-well depth, V_0, with the range, b, for the binding energy $B_d = 2.225$ MeV, and the variation of V_0 with b for the three values of $R_{rms} = 2.10 \pm 0.06$ fm, the root-mean-square radius of the electric charge distribution. The intersection for $R_{rms} = 2.10$ fm yields $V_0 = 34.6$ MeV and $b = 2.07$ fm, with b fortuitously close to R_{rms}. (Burge 1986.)

different from those for the hydrogen atom but they require extra care in interpretation since the coordinate r is the internucleon distance and not the distance from the centre-of-mass. For the hydrogen atom the centre-of-mass is very close to the centre of the proton, but for the deuteron it is almost exactly midway between the neutron and proton. The probability per unit volume of finding the proton and neutron separated by a distance r is equal to the probability per unit volume of finding the proton (and the neutron) a distance $r/2$ from the centre-of-mass. This corresponds to the electric charge density measured in electron scattering when allowance is made for the finite size of the proton's charge distribution.

The analysis of this approach to the deuteron can be extended to investigate the possibility of excited states of the deuteron (see

Appendix C). It is found that the first excited state is unbound by about 200 MeV for the preferred values of V_0 and b, in agreement with the fact that no excited state of the deuteron has been observed in experiments.

More detailed studies with a variety of reasonable well-shapes yield similar relationships between a characteristic well-depth parameter and a characteristic range, and also exclude excited states. Other available experimental evidence, such as the deuteron spin ($J = 1$) and magnetic moment, indicates that the intrinsic spins ($s = \frac{1}{2}$) of the proton and neutron are parallel and the orbital angular momentum is zero ($l = 0$), confirming the reasonableness of the assumption of spherical symmetry for the wavefunction. (The finite but small electric quadrupole moment of the deuteron indicates that the charge distribution is not exactly spherical. A small admixture of an $l = 2$ state is required and this is evidence of non-central forces such as can be provided by a tensor component in the potential, see p. 124.)

We must now see whether the square-well potential with the deduced parameters, $b \simeq 2.1$ fm and $V_0 \simeq 34$ MeV, can also explain low-energy neutron–proton scattering.

Neutron–proton scattering

The quantum-mechanical theory of scattering, even in its simplest form, is beyond the scope of this book. A few general results, both experimental and theoretical, can be usefully presented since they lead to an important advance, conceptually simple, in our understanding of nuclear forces.

For neutron energies below a few MeV it is found that the angular distribution of the differential cross-section $d\sigma(\theta)/d\Omega$ for scattering by protons is spherically symmetric. This is characteristic of the simplest solution of the wave equation for which the (quantized) angular momentum of the neutron about the proton is zero (classically this would imply head-on collisions), and in an oversimple interpretation of quantum mechanics it implies that the product of the linear momentum and the impact parameter (Appendix A, Fig. A.1) is less than about $\hbar$. Since our deuteron potential was derived for an $l = 0$ state we are encouraged to see how well it predicts the magnitude of the observed total elastic cross-section ($= 4\pi d\sigma(\theta)/d\Omega$). Theory gives, in good approximation,

$$\sigma_{\text{el}} (l = 0) = 5.2/(T_{\text{n}}(\text{CM}) + B_{\text{d}}) \text{ barn,}$$

with B_d, and also the neutron energy in centre-of-mass coordinates $(T_n(CM) \simeq \frac{1}{2}T_n(LAB))$, expressed in MeV. It is clear that this indicates, since $B_d = 2.23$ MeV, that σ_{el} is closely independent of energy for $T_n \lesssim 0.1$ MeV. This independence is in excellent agreement with experiment but, unfortunately, the magnitudes do not agree:

$$\sigma_{el} \text{ (theory)} \simeq 2.3 \text{ b}, \quad \sigma_{el}(\text{experiment}) \simeq 20.4 \text{ b}.$$

It was Wigner in 1935 (unpublished) who first proposed a satisfactory solution. For the deuteron the proton and neutron spins are parallel, but for a collision between a neutron and a proton the spins can be either parallel or antiparallel. For $l = 0$ the total spin is then either $J = 1$ or $J = 0$. The former is referred to as a triplet state since J can take the values $+1$, 0, and -1 in relation to a defined direction (e.g. the perpendicular to the scattering plane), and the latter is a singlet state $(J = 0)$. In scattering the triplet state will be three times as probable as the singlet, and we can expect

$$\sigma_{el} = \tfrac{3}{4} \sigma_{el}(\text{triplet}) + \tfrac{1}{4}\sigma_{el}(\text{singlet}).$$

from which it follows that σ_{el} (singlet) needs to be 70.6 b in order to account for σ_{el}. The only way to achieve this is to adopt a singlet potential with V_0 or b, or both, different from the triplet potential of the deuteron. With help from other experiments, e.g. the n(p,γ)d cross-section, the values deduced are:

Square well	Triplet	Singlet
Depth V_0	38.5 MeV	14.3 MeV
Range b	1.93 fm	2.50 fm

The conclusion is inevitable; the nuclear interaction is, in general, spin-dependent. The singlet state is unbound by about 50 keV (since its V_0 is relatively small, with b insufficiently larger) and does not enter the deuteron problem. This spin-dependence is not to be confused with spin-orbit forces (see p. 132). The latter vanish for $l = 0$, i.e. zero orbital angular momentum. The former are clearly still present since all our discussion has been for $l = 0$. Both spin-dependent forces and spin-orbit forces require a further tensor term in the potential (see p. 123). A familiar classical system requiring a tensor force is two interacting magnetic dipoles, since the force obviously depends on their relative orientation. (See G. A. Jones, *The structure of nuclei*, OPS12.)

To obtain further information about the neutron–proton interaction it is necessary to study scattering at higher energies, together with more sophisticated conditions, e.g. aligning the spins of the incident particle (or the target) in polarization experiments.

Yukawa's field quantum—the pion

The inverse-square law for electrostatics has been accounted for in the elegant but difficult and controversial theory of quantum electrodynamics. This is based on the assumption that the interaction is mediated by photons exchanged between the electric charges. The photons have energy $h\nu$, and there might seem to be, at first sight, a failure to conserve the energy of the system. However, Heisenberg's principle allows an uncertainty of energy ΔE within the time interval Δt provided that $\Delta E \Delta t \lesssim \hbar$. For an interaction at distance r, mediated at the velocity of light c, we can write $\Delta t \simeq r/c$ and since $\Delta E = h\nu$ the wavelength of the ('virtual') photon is $\lambda = 2\pi r$.

Yukawa (1935) realized that 'in the quantum theory [the nuclear] field should be accompanied by a new sort of quantum', and he was cautious enough to add, 'besides such an exchange force and the ordinary electric and magnetic forces there may be other forces between the elementary particles.' He derived an expression for the nuclear potential $V = (-g^2/r) \exp(-kr)$ (Fig. 7.5) and estimated the two constants g and k by comparison with experiment, concluding

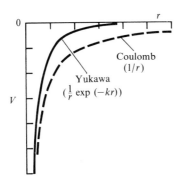

Fig. 7.5. Comparison of the form of the Yukawa 'one-pion exchange' potential with the form of the Coulomb potential.

that $k = 10^{14} - 10^{15} \, \text{m}^{-1}$ and 'g a few times the elementary charge e'. It is now common to compare the so-called 'coupling constants' $g^2/hc \simeq 1$ for 'strong' nuclear and $(1/4\pi\varepsilon_0)(e^2/hc) \simeq 1/137$ for electromagnetic forces.

Yukawa considered the 'nuclear field quantum' to have an electric charge $+e$ or $-e$, and he deduced that the rest mass M of this particle must be about 200 times the electron mass m_e. This followed from the fact that nucleon–nucleon forces act strongly over a short range, $R = \sim 2 \, \text{fm}$, i.e. about the radius of the deuteron. Thus we can repeat the consideration of Heisenberg's uncertainty relation with $\Delta E = Mc^2$ and $\Delta t \simeq R/c$, yielding $M \simeq \hbar/Rc$. For $R = 2 \, \text{fm}$, $M \simeq 200m_e$. With a humble over-confidence in the pre-eminence of experimentalists Yukawa remarked, 'as such a quantum with large mass and positive or negative charge has never been found by the experiment, the above theory seems to be on a wrong line.'

Two years later, with the discovery of the muon possessing just these characteristics $(m_\mu \simeq 207m_e, \; q = \pm e)$, Neddermeyer and Anderson (1937) seemed to have removed Yukawa's doubts. Unfortunately, the muon did not react at all strongly with protons or other nuclei, and it was not until 1947 when Powell and co-workers observed the pi-mesons (or pions) in nuclear emulsions exposed to cosmic rays from outer space that the identity of Yukawa's particle became established $(m_\pi \simeq 273m_e, \; q = \pm e$, and $m_\pi \simeq 264m_e$ for $q = 0)$. The charged meson (Lattes, Occhialini and Powell 1947) decays to form a muon, which in turn decays to form an electron. Perkins (1947) first observed the 'explosive' interaction of a pion with a nucleus, and within a few years many new unstable particles were observed (see Chapter 8).

The simple 'one-pion exchange potential' of Yukawa is useful as a guide, especially for nuclear forces beyond about 2 fm, but has a number of unsatisfactory features. For example, it is now generally agreed that some kind of repulsive core potential is required at a range below about 0.5 fm in order to explain high-energy nucleon–nucleon scattering, and such a core helps to explain the 'saturation' of nuclear forces revealed by the approximately constant density of nuclear matter. Early meson theories with certain types of exchange force produced the repulsive terms required to prevent the collapse of nuclear matter to nucleon dimensions, but they could not explain the high-energy scattering. Particles other than pions (e.g. kaons, see p. 146) are also known to interact strongly with nucleons, and to explain polarization effects during nuclear scattering vector mesons with non-zero spin are required, yet pions and kaons have zero spin. It may be that the full nucleon–nucleon interaction is a problem of complexity rather than of principle, with the quark–gluon structure of nucleons increasingly important at higher energies (see Chapter 9).

The weak interaction—beta-emission

One important type of 'nuclear' interaction not covered by the forces discussed so far is the emission of electrons and neutrinos by nuclei (see pp. 19, 22), of which the simplest example is the beta-decay of the free neutron

$$n \rightarrow p + e^- + \bar{\nu}_e + Q \quad (\tau_{\frac{1}{2}} = 10.8 \text{ min}, \ Q = 0.78 \text{ MeV}).$$

The probability of these decays, i.e. the reciprocal of the mean life τ_{mean}, is determined by the interaction potential involved. The strong nuclear potential would produce beta-decay lifetimes that are much too short, indicating that it is much too strong. Fermi (1934) developed a theory of beta-emission that successfully accounted for the continuous energy spectrum of the observed beta-particles (see G. A. Jones, OPS 12). The theory also yielded expressions for the half-life in terms of the maximum energy of the betas and other parameters. One important parameter was the strength or 'coupling constant' of this so-called weak (nuclear) interaction which turned out to be about 10^{-13} compared with about unity for the strong interaction. Yukawa in 1938 suggested there should be a field quantum for the weak interaction, corresponding to the photon for the electromagnetic interaction. In the 1960s this idea was resurrected and the corresponding 'intermediate vector bosons', the W^+, W^-, and Z° particles, were first detected as free particles in 1983 by teams at CERN (Geneva) in colliding beam experiments using 270 GeV protons and antiprotons (Chapter 9). Other examples of the weak interaction occur in the decays of the muon (see p. 144), of the pion and other mesons, and of hyperons, and also in electron capture (see p. 97).

Summary of types of interactions

In Table 7.1 the well-established types of interactions are summarized. The electromagnetic and weak interactions are now known to be

TABLE 7.1

Interaction	Source	Field quantum			Strength or coupling constant
		Particle	Rest mass	Spin	
Strong nuclear	BARYONS Nucleons, hyperons, baryon resonances	MESONS Pions, kaons, etc., meson resonances	Finite, 135 MeV/c^2 to over 2.5 GeV/c^2	0, 1, 2, (3)	$\simeq 1$ to ~ 15
	HADRONS				
Strong interquark	QUARKS	GLUONS	0	1	—
Electro-magnetic	Electric charges	Photon	0	1	$\simeq 10^{-2}$
Weak nuclear	LEPTONS Electrons, muons, tauons, and their neutrinos	Inter-mediate vector bosons, W$^\pm$ & Z^0 particles	Finite, W$^\pm$ = 82.3 GeV/c^2 Z^0 = 92.8 GeV/c^2	1	$\simeq 10^{-13}$ to $\sim 10^{-5}$
Gravity	Masses	(Graviton)	(0)	(2)	$\simeq 10^{-38}$

Bracketed entries have been postulated or deduced from theory, but have not yet been identified.

different manifestations of a single united 'electroweak' force. Attempts continue to unify all four interactions (Chapter 9).

We note that the electromagnetic interaction takes place between bodies or constituents of matter with electric charge which act as 'sources' of the electromagnetic field quantum, or force carrier, the photon, with zero rest mass and spin 1. Gravity acts between bodies with mass ('source particles'), and the hitherto undetected 'graviton' has been postulated as the field quantum to mediate the interaction, with zero rest mass, in accord with the inverse-square law. Theory beyond the level of this treatment suggests a graviton spin of 2. The strong nuclear interaction takes place between 'source' particles collectively known as 'baryons' (meaning 'heavy ones'), Pais (1953), which include neutrons and protons and 'hyperons' (meaning the mass is 'greater than' the nucleon mass) and 'baryon resonances' (Chapter 8). The field quanta are 'mesons' (meaning mass is 'between' those of the electron and proton, although later discoveries included mesons with masses greater than that of the proton), e.g. pions, kaons, which also appear to combine to form more complex unstable particles ('meson resonances', Chapter 8). The particles which interact with each other through the strong nuclear interaction, i.e. baryons and mesons, are known as 'hadrons' (meaning 'strong ones'). Now that it is evident that hadrons consist of quarks (source particles) held together by gluons (field quanta), a 'strong interquark force' is identified (Chapter 9). This may be seen as the nuclear equivalent of the electromagnetic force binding electrons to nuclei in atoms. The 'strong nuclear force' between nucleons and nuclei would then be the nuclear equivalent of the electromagnetic forces (Van der Waals forces) between atoms and molecules. The sources of the weak interaction are leptons, or more accurately 'electron leptons' for $e^{\pm}$, v_e, $\bar{v}_e$, 'muon leptons' for $\mu^{\pm}$, v_μ, $\bar{v}_\mu$ and 'tauon leptons' for $\tau^{\pm}$, v_τ, $\bar{v}_\tau$ (Chapter 9)—the smallest Greek coin, the widow's mite of the New Testament, was a lepton, meaning 'the small one'.

The coupling constants given for the electromagnetic and gravitational interactions are related to field quanta of zero mass. For the other interactions, which have finite mass field quanta, the coupling is dependent on factors such as the energy of the interaction or the distance between the source particles. Thus, for β-decay energies of about 1 MeV the corresponding coupling constant value $\sim 10^{-13}$ is consistent, but at GeV energies a value of $\sim 10^{-5}$ is deduced. Similarly, for the strong interaction a value of 15 is given for GeV energies. For the strong interquark interaction there is reason to believe that point-like quarks at very short separations experience a negligible interaction (so-called 'asymptotic freedom') but as the separation increases to ~ 1 fm the energy of separation becomes larger than the rest mass of the pion, and then pions are preferentially formed as more energy is supplied, rather than free, isolated quarks, which have not (yet) been observed. The inability to free quarks at long distances (long wavelengths) is referred to as 'infrared slavery'.

Nuclear models

The application of quantum electrodynamics to the hydrogen atom has been entirely successful within the very narrow limits set by precision measurements of the energy levels and relative intensities of spectral lines (i.e. transition probabilities). Atoms containing many electrons have a dominant centre of force, the nucleus, but residual interactions between the electrons introduce problems that can be solved only approximately. For nuclei there is no obvious centre of force. The many nucleons interact with all their neighbours and maybe with nucleons at a slightly longer range; to make matters worse, the exact form of the nucleon–nucleon interaction is not known. In such circumstances the physicist first resorts to models. Brief mention has been made of the optical model (see p. 42f, 100) which accounts for the scattering and absorption of incident particles in terms of a 'cloudy crystal ball'. In practice, this means inserting into the wave equation an equivalent potential with real and imaginary parts (the latter to simulate absorption) and a special spin-orbit term to yield polarization effects (see p. 132).

The shell model

We have already considered the bound and unbound states of the neutron–proton system in terms of a simple square-well potential. Surprisingly, there was no significant success in the attempt to consider the properties of nucleons in larger nuclei in terms of suitable potentials until 1948. The possibility of nucleon 'orbits' within the dense nuclear matter was considered to be negligible. However, from accumulated evidence it became abundantly clear that nucleons occupy well-defined energy states and certain numbers of nucleons lead to extra-stable nuclei. This is reminiscent of electrons in atomic shells where the shell 'closures' are at 2, 10, 18, 36 electrons, although the higher numbers are obscured by the next shell beginning to be populated before the previous shell is completely filled, depending on which has the lower energy for the next electron Fig. 7.6(a). The corresponding numbers for *either* neutrons *or* protons are 2, 8, 20, 28, 50, 82, and 126 and are known as 'magic numbers'. In Fig. 7.6 (a)–(e) are shown a few examples of marked changes in nuclear properties at these numbers, including for comparison one example of periodic properties in atoms (see also Fig. 7.8).

A simple harmonic oscillator (parabolic) potential yields energy levels (Fig. 7.7(a)) corresponding to 2, 8, 20, 40, 70, 112, etc. neutrons or protons—encouraging but not good enough. Other reasonable shapes split some of these levels to produce for example, 2, 8, 18, 20, 34, 40, and 58 (Fig. 7.7(b)), but none could produce 28 and 50 until

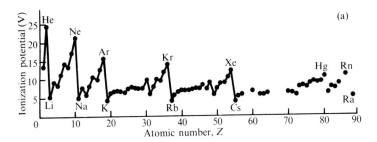

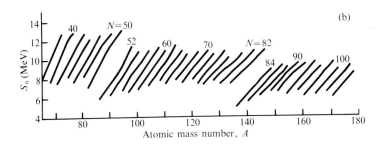

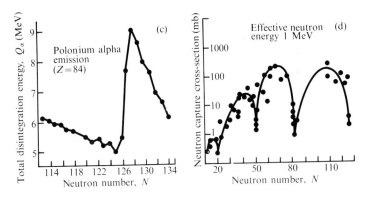

Fig. 7.6. Evidence for 'magic' numbers. (a) First ionization potentials (electron separation energies) of *atoms* indicating filled electron shells at $Z = 2, 10, 18, 36, 54$. (b) Neutron separation energies for even-N nuclei as function of A (odd and even) showing drops at $N = 50, 82$ (cf. (a)). (c) Disintegration energies for alpha-emission from polonium isotopes showing relatively high Q_α for $N = 128$, indicating ease of removal of two neutrons outside $N = 126$ 'core'. (d) Neutron capture cross-sections for fission neutrons of effective energy 1 MeV showing very low values (note logarithmic scale) for $N = 20, 50, 82, 126$.

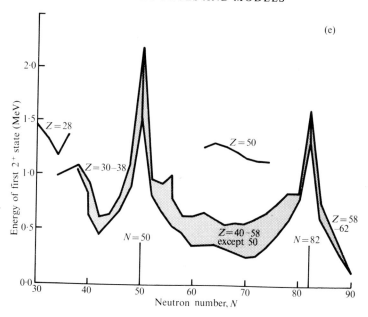

Fig. 7.6. (*cont.*) (e) Energies of the first 2^+ states of even–even nuclei showing high values for magic numbers of protons (28, 50) and of neutrons (50, 82).

Mayer (1948), and independently Haxel, Jensen, and Suess (1949), introduced a term in the potential that involved the spin of the particles. More accurately, it related the particle spin (σ) to its orbital angular momentum (**l**), and by suitably adjusting the strength V_s in the 'spin-orbit' term $V_s f(r)\sigma.\mathbf{l}$ it was possible to split the levels even further and also produce energy gaps in the sequence of levels that corresponded exactly to the magic numbers (Fig. 7.7(c)). (A similar spin-orbit term in atomic structure theory is much weaker and produces very small splitting of atomic levels.)

When such a spin-orbit term is included in the nuclear optical potential (see p. 42) it leads to different probabilities of scatter to left and right for an incident beam that is polarized, i.e. a beam of particles having a different probability of their spin vector σ pointing 'up' with reference to the scattering plane rather than 'down'. It follows that the sign of the spin-orbit potential term, $V_s f(r)\sigma \cdot \mathbf{l}$, is positive or negative depending on the direction of the **l** vector, i.e. on whether the impact parameter relates to an incident particle on the right or left of the nucleus.

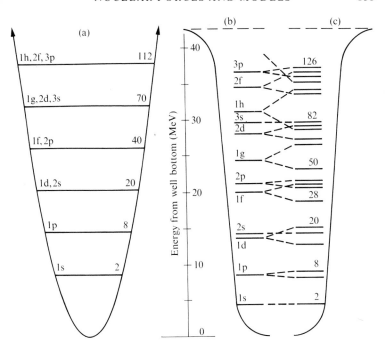

Fig. 7.7(a) Nuclear energy levels for an harmonic-oscillator potential showing the degeneracy (i.e. equal energies) of some levels (1d, 2s), (1f, 2p), etc. The nomenclature s, p, d, f, g, h corresponds to $l = 0, 1, 2, 3, 4, 5$, and the numbers of particles in each state is given by $2(2l + 1)$. (b) Energy levels for a potential of the shape shown with $R \sim 10$ fm and no spin-orbit term. The degeneracy of (a) has been removed and significant energy gaps occur at proton (or neutron) numbers 2, 8, 20, 34(1f), 40, 58(1g), 68(2d), 70(3s), 92(1h), 106(2f), 112(3p). (c) Effect of spin-orbit term on levels of (b). Apart from the s-levels, each level is split into two, the lower with $j = l + \frac{1}{2}$, the upper with $j = l - \frac{1}{2}$. The magnitude of the splitting increases with l and the magic numbers can be obtained as shown, e.g. 28 is produced by the $1f_{7/2}$ state (with $j = \frac{7}{2}$) with 8 particles beyond the 20 in the lower shells.

This nuclear 'shell model' also accounted for many, but not all, of the spins of the ground states of nuclei, but was much less successful for magnetic moments. It failed almost completely to explain the numerous excited states of nuclei, and these and other properties were soon recognized to require a departure from the simplifying assumption that nuclei are spherical.

The collective model

Aage Bohr and Mottelson (1953) investigated the consequences of collective motion of nucleons which produces ellipsoidal oblate and prolate nuclei. In particular, they studied the energy levels of such nuclei when they rotate and vibrate. The results are particularly simple for pure (quadrupole) vibration, yielding equally spaced levels,

$$E_{vib} = (n + 5/2)\hbar\omega,$$

where $n = 0$, 1, 2, etc., and ω is a characteristic frequency. For pure rotation the relation is

$$E_{rot} = R(R + 1)\hbar^2/\mathscr{I}_{eff},$$

where $R = 0, 2, 4$, etc., and $\mathscr{I}_{eff}$ is the effective moment of inertia (less than that for a rigid body). A test of these relationships is provided by taking the ratio of the energies of the excited states to that of the first state. We expect 3.33, 7, 12, and 18.33 for rotational states and find, for example, for $^{164}_{68}$Er (Erbium), 3.27, 6.7, 11.2, and 16.0. In Fig. 7.8 we show the ratio for the first two excited states for many even-Z–even-N (even–even) nuclei and the value 3.33 expected for 'pure' rotational states is clearly shown. For the equally spaced levels of vibrational states this ratio is expected to be 2, but this is not so clearly or frequently exhibited, since in general rotational and vibrational motion is mixed or 'coupled' and the results can be rather complicated.

The purer vibrational states occur for near-spherical nuclei, found near magic numbers of Z or N, and are widely spaced ($\lesssim 1$ MeV). Rotational states are more evident for distorted nuclei that occur away from magic numbers, and usually have spacings nearer to ~ 100 keV.

Studies with collisions of heavy ions indicate that nuclei can exist with more complicated shapes such as ellipsoids with three different axes, and also dumb-bell shapes.

The liquid-drop model and the semi-empirical mass formula

As early as 1935, von Weizsäcker developed a formula for the masses of nuclei, and hence their binding energies, based on the observation that the density of nuclear matter is essentially constant (see pp. 42 and 43), i.e. rather like a liquid (a model probably first suggested by

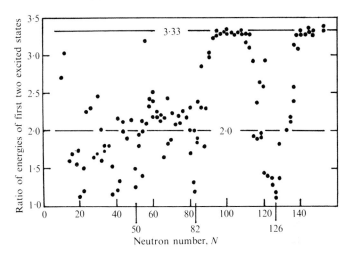

Fig. 7.8. Evidence for rotational energy levels in the non-spherical even–even nuclei with neutron numbers away from magic numbers.

Gamow 1930). This 'liquid-drop model' suggests binding energies expressible in terms of 'latent heat of vaporization' or volume energy and 'surface energy', together with allowances for the Coulomb repulsion of the protons. The dominant terms are

$$B = m_pZ + m_nN - m(A, Z) = a_1A - a_2A^{\frac{2}{3}} - a_3Z^2A^{-\frac{1}{3}},$$

binding energy	proton masses	neutron masses	mass of nucleus	volume energy	surface energy	Coulomb repulsion

where a_1, a_2, and a_3 are constants, with a_1 corresponding to the binding energy per nucleon for infinite uncharged nuclear matter, and a_2, a_3 reducing the binding for finite nuclei. Another term, $-a_4(A -2Z)^2/A$ allows for a diminution of binding related to $A-2Z$, the excess of N over Z in most nuclei, and yet another $-a_5(A, Z)$ for the fact that, in nuclei, pairs of protons (or neutrons) are in a state of lower energy when their spins are antiparallel. Thus, a_5 is negative for A even, Z even, zero for A odd, and positive for A even, Z odd. With only five adjustable constants, the agreement with the measured binding energies of hundreds of nuclei is remarkably good (Fig. 7.9, cf. Fig. 6.5).

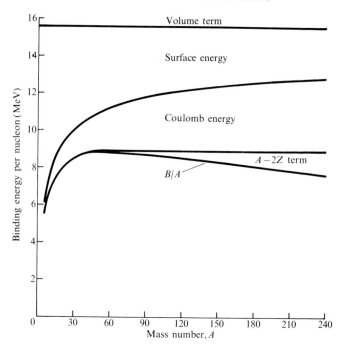

Fig. 7.9. Importance of the four principal terms in the semi-empirical mass formula illustrated by the binding energy per nucleon, $B/A = a_1 - a_2 A^{-1/3} - a_3 Z^2 A^{-4/3} - a_4(A - 2Z)^2 A^{-2}$ with $a_1 = 15.6$, $a_2 = 17.2$, $a_3 = 0.70$, and $a_4 = 23.3$ MeV. Notice the remarkable (but fortuitous) constancy of B/A for the sum of the first three terms, for $A > 45$, and the relatively small effect of the neutron-excess term $A - 2Z$.

The liquid-drop model has been applied to explain aspects of nuclear reaction mechanisms, notably by Niels Bohr (1936) in his idea of the compound nucleus (see pp. 52, 99, 106) and by Meitner and Frisch (1939) in their recognition of the possibilities and consequences of nuclear fission. The changes of surface energy and Coulomb energy during deformation can be shown to lead to fission provided Z^2/A is sufficiently large, as for the heaviest nuclei, and the sequence of changes of shape through elongation, necking, and break-up are not difficult to imagine.

Problems

7.1. The Schrödinger equation for a particle of mass μ in a central potential $V(r)$ is $\nabla_r{}^2 \psi(r) + (2\mu/\hbar^2) \{E - V(r)\} \psi(r) = 0$

and $\nabla_r^2 \psi = \dfrac{1}{r^2} \dfrac{\partial}{\partial r}\left(r^2 \dfrac{\partial \psi}{\partial r} \right) + \dfrac{1}{r^2 \sin \varphi} \dfrac{\partial}{\partial \varphi}\left(\sin \varphi \dfrac{\partial \psi}{\partial \varphi} \right) + \dfrac{1}{r^2 \sin \varphi} \dfrac{\partial^2 \psi}{\partial \theta^2}.$

Show that a spherically symmetric function $\psi = U(r)/r$ allows the wave equation to reduce to the simple form

$$\frac{d^2 U(r)}{dr^2} + \frac{2\mu}{\hbar^2}\{E - V(r)\}\ U(r) = 0,$$

as used for the deuteron.

7.2. Show that the matching of the magnitudes U and slopes dU/dr of the wavefunctions for the deuteron $U(r < b)$ and $U(r > b)$ at $r = b$ leads to $K \cot Kb = -\alpha$ (see p. 189). Hence show that the solution is near to $Kb = \pi/2$ and that this value for Kb leads to $V_0 b^2 \simeq \pi^2 h^2/8\mu = 1.0$ MeV b.

7.3. The deuteron continuity relation can be written $\tan Kb = -Kb/\alpha b$. Show that $\alpha b = 0.487$ for $B_d = 2.23$ MeV and $b = 2.1$ fm. Solve for Kb by plotting $\tan Kb$ and $-Kb/0.487$, hence deduce the corresponding values of V_0 and $V_0 b^2$.

7.4. Calculate the impact parameter for 14 MeV neutrons on protons such that the total orbital angular momentum is $\hbar$. Compare this with the ranges of the square-well potentials used to account for neutron–proton scattering, and comment on the likelihood of 14 MeV neutron–proton scattering being spherically symmetric.

7.5. The classical formula for the Coulomb energy of a nucleus is $(1/4\pi\varepsilon_0)3Z(Z-1)e^2/5R$. The optimum value of the coefficient a_3 of the corresponding term $a_3 Z^2 A^{-1/3}$ in the semi-empirical mass formula is found to be 0.70 MeV/c^2. Compare this value with the theoretical expression above, using $R = 1.2\ A^{1/3}$ fm.

7.6. Calculate the total binding energy B and the binding energy per nucleon B/A for $^{208}_{82}\text{Pb}$. Express B in terms of the rest-mass energy of a proton. What fraction of B is surface energy? (See caption to Fig. 7.9. Use the atomic masses for ^{1}H, 1.007825 u, and ^{208}Pb, 207.976650 u, and the neutron mass, 1.008665 u.)

7.7. Use the semi-empirical mass formula (see p. 135 and Fig. 7.9) to estimate the expected total disintegration energy in alpha-emission for $^{197}_{84}\text{Po}$, $^{204}_{84}\text{Po}$, $^{209}_{84}\text{Po}$, $^{213}_{84}\text{Po}$, and compare with the values in Fig. 7.6(c). The binding energy of the alpha-particle is 28.3 MeV.

7.8. Derive an expression for the atomic number of the most stable member of a set of isobars, using the semi-empirical mass formula including the neutron-excess term (Fig. 7.9) but not the pairing-energy term. Deduce the most stable members for $A = 17$, $A = 41$, and $A = 79$, and comment on the neglected factors that affect stability $(m_n - m_p = 1.3 \text{ MeV}/c^2)$.

8. Cosmic rays and strange particles

Introduction

The discovery of radioactivity prompted the development of sensitive instruments to measure the ionization produced in air by the alpha, beta, and gamma rays. These studies indicated that ionization could not be altogether removed by shielding or by careful purification of the gas and other materials. The same instruments were installed in balloons, and registered greatly increased ionization at altitudes up to about 16 000 m. It was concluded by Hess (1913) that the effect was due to ionizing radiation from outside the Earth, and Millikan and Bowen (1926) suggested the name 'cosmic rays'.

Cosmic rays

At first it was commonly assumed that the cosmic rays were gamma-rays, but sea-level studies (during a cruise from Europe to Java) indicated a systematic decrease with decreasing latitude, which was most readily interpreted as due to the Earth's magnetic field acting on positively charged particles of energies up to a few GeV, the higher-energy particles being relatively unaffected.

A great deal of effort has been expended in determinations of the distribution of the particles in space and time, their nature, and their energy spectra. In general the primary rays, outside the Earth's atmosphere, are nuclei with abundances very close to those of the elements in the universe, but with markedly more lithium, beryllium, and boron. Some 94 per cent are protons, and 5 per cent are ^{4}He and ^{3}He. It is probable that the lithium, beryllium, and boron nuclei result from proton disintegrations of intergalactic matter. Small percentages of electrons and gamma-rays are also reported. The number with energy E in the interval dE is proportional to $E^{-\gamma}dE$, with $\gamma \simeq 1.5 - 2.0$ in the range $E = 1 - 10^{10}$ GeV and is about 10^4 m^{-2} s^{-1} sr^{-1} at $E = 1$ GeV, for $dE = 1$ MeV. Any theory of the origin of the cosmic rays must explain this energy distribution, together with the fact that their intensity is remarkably constant and isotropic, i.e. independent of time and of the direction in space. These are the dominant features of cosmic rays, but there are also fluctuations which can be explained in terms of rays from the Sun, and possibly, in the case of gamma-rays, by individual stellar locations.

We shall not pursue these interesting astrophysical and cosmological topics but confine ourselves to those aspects that advanced our understanding of nuclear physics. They arose almost entirely from the interaction of the high-energy primary cosmic-ray protons with nuclei in the atmosphere. A wide variety of particles is produced, including the pions (and kaons) already discussed, their decay products (muons and neutrinos), and high-energy gammas that produce positron–electron pairs.

Positrons and antiparticles

The introduction of the positron is part of a story with a moral. A relativistic generalization of the wave equation suitable for charged particles in an electromagnetic field was first developed by Schrödinger (although it is now known as the Klein–Gordon equation), and was almost immediately rejected because it led to negative as well as positive total energies for the particles,

$$E = \pm (p^2c^2 + m_0^2 c^4)^{\frac{1}{2}},$$

where m_0 is the rest mass and p the momentum. Dirac (1928) developed a more complete equation which not only yielded the same result for the energy states but also automatically predicted an intrinsic angular momentum or spin of $\frac{1}{2}\hbar$ for the particle and the correct magnetic moment. The Klein–Gordon equation appears to be valid for particles with zero intrinsic spin. The possible energy states are shown schematically in Fig. 8.1, and the positive ones are precisely those expected for ordinary particles, e.g. electrons, with kinetic energy $T = E - m_0 c^2$.

The negative energy states suggest that transitions could occur from the positive energies with the emission of electromagnetic energy, i.e. photons. But such transitions had not been observed, therefore all the negative states were boldly considered by Dirac to be already occupied. (Lest this should worry readers by implying that the total energy of the universe is negative, remember that the zero of energy is arbitrary—only changes of energy can be measured.) This

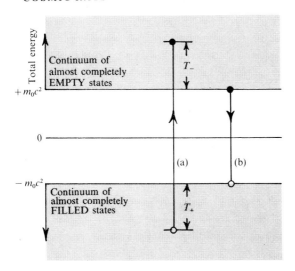

Fig. 8.1. Energy states for free electrons. The transition (a) corresponds to pair production of a particle (e.g. an electron) of kinetic energy T_- and of an antiparticle (e.g. a positron) of kinetic energy T_+. The transition (b) corresponds to positronium annihilation.

suggests, however, that the supplying of sufficiently energetic photons to a suitable system would promote an electron from a negative energy state to a positive one, thereby making it observable as an ordinary electron. The unfilled negative energy state, or 'electron hole', would then have the properties of a 'positively charged electron'. At the time (1928) such a particle was not known and Dirac thought at first that it might be the proton, in spite of the discrepancy in rest mass.

Four years later, Anderson (1932) observed an electron-like track in a cloud-chamber photograph (Fig. 8.2(a)), with clear evidence that the particle was positively charged. Its direction of motion was given by its smaller radius of curvature (produced by a magnetic field) after passing through a metal plate, and its direction of curvature then indicated the positive charge. The identification of the 'positron' of Dirac's theory was obvious but not yet complete. Electron–positron pair production by gammas of sufficient energy ($h\nu > 2\,m_e c^2 = 1.022$ MeV) was soon observed in cloud chambers (Fig. 8.2(b)), the interaction having taken place between the photon and the very strong electric field near a nucleus. Finally, the 'annihilation' of a

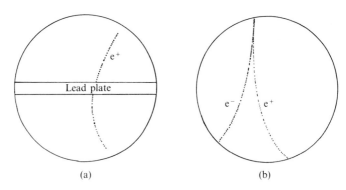

Fig. 8.2. Cloud-chamber pictures of positrons. (a) From the photograph that led to Anderson's discovery of the positron. (b) An example of electron–positron pair production in the wall of the cloud chamber (the applied magnetic fields were perpendicular to the plane of the diagram).

positron by combination, at rest, with an electron was observed from the two photons, each of characteristic energy 0.511 MeV, which accounted for the rest-mass energy of the two particles. It has since been found that the electron–positron system forms a kind of 'electron atom' known as positronium, which with antiparallel spins has a mean life of about 8 ns. There is also a triplet state with parallel spins and zero orbital angular momentum and a mean life of about 7 μs, forming three photons with total energy 1.022 MeV.

There is nothing in Dirac's approach to the relativistic wave equation that restricts the results to electrons. Other particles, such as the proton and neutron, can be expected to have negative energy states, but it would require gammas of energy greater than 2 × 938 MeV to produce a proton pair, or 'proton–antiproton' as they are called. Such high energies are now readily available by means of accelerators, and it is even possible to produce beams of antiprotons, admittedly of relatively low intensity, for studying their properties in scattering and reactions, including 'proton–antiproton annihilation' which leads to the production of pions, and kaons, and at high collision energies to a whole range of unusual particles. The antiproton is negatively charged, and was first identified in 1955 from proton bombardment of copper nuclei at energies above 4.3 GeV. For proton–proton collisions the threshold is $T_p = 5.6$ GeV, but protons in the copper nuclei can be moving towards the incident protons. A mere 30 MeV of target proton energy can reduce the threshold to about 4.3 GeV. The proton is now known to consist of three quarks, bound by gluons (see p. 164), and the antiproton has three antiquarks.

The antiparticle of the neutron has also been observed. It has zero charge and, as for all antiparticles, its mass and spin are identical with those of the corresponding particle, but the magnetic moment of the antineutron is of opposite sign to that of the neutron in relation to the direction of their spins.

Mesons also have their antiparticles with equal masses and (integral) spins but opposite electric charges and magnetic moments. They are not necessarily created or annihilated in pairs.

Particles that have no electromagnetic interaction sometimes have antiparticles, e.g. the neutrinos (v_e, v_μ, v_τ), and sometimes they are 'self-conjugate', i.e. particle and antiparticle are experimentally indistinguishable, as for the neutral pion (π^0).

Muons, tauons, and lepton numbers

The discovery in 1937 of the muon of mass about 200 times that of the electron appeared at first to support Yukawa's prediction of a nuclear field 'quantum' (see p. 126). The particle was observed as a curved track originating in a lead plate in a cloud chamber operated on a high mountain (Pike's Peak in Colorado), and its droplet density and curvature (i.e. magnetic rigidity, $BR = mv/e$) together indicated a mass between that of an electron and a proton. Anderson and Neddermeyer proposed the name mesotron, which was later abbreviated to meson (Greek: *meson* = thing in the middle or in between); later still, this was sharpened to mu-meson when the pi-meson was discovered, and yet later abbreviated to muon to prevent confusion when the term meson became restricted to the field quanta (pions, kaons, etc.) of the strong nuclear interaction. The fact that the muon did not interact strongly with nuclei ruled out the possibility of its being Yukawa's particle—the pion was to fill that role. It was later found that the muon decayed, with a mean life of 2.2 μs, to an electron or positron, depending on the muon charge; and the continuous energy spectrum of the electrons (similar to that in beta emission) indicated the presence of (at least) two other particles. Careful mass and energy measurements supported the possibility of two neutrinos, and in the early 1970s it became increasingly evident that these are different types of neutrino. One type, the electron neutrino v_e, can interact with nuclei to produce electrons; the other type is able to produce only muons, and is known as the muon neutrino v_μ. A third type of lepton was discovered by Perl and co-workers (1975) with a mass of $1.8\ \mathrm{GeV}/c^2$, i.e. over 3500 times the mass of the electron. It is known as the tauon, τ, and indirect evidence shows that there is a corresponding tauon neutrino, v_τ.

Neutrinos, like other particles, have their corresponding antiparticles, and it is found that the creation and annihilation of electrons and positrons in pairs is a special case of the creation and annihilation of leptons in pairs. This has been represented in a helpful way by the introduction of lepton numbers of three types, one for electrons and their neutrinos (L_e), another for muons and muon neutrinos (L_μ), and the third for tauons and tauon neutrinos (L_τ), as shown below:

	e^-, v_e	$e^+, \bar{v}_e$	μ^-, v_μ	$\mu^+, \bar{v}_\mu$	τ^-, v_τ	$\tau^+, \bar{v}_\tau$
L_e	$+1$	-1	0	0	0	0
L_μ	0	0	$+1$	-1	0	0
L_τ	0	0	0	0	$+1$	-1

The sum total of L_e is found to be conserved in any interaction, strong, weak, or electromagnetic. The same is true for L_μ and L_τ.

Thus
$$\mathrm{n} \to \mathrm{p} + \mathrm{e}^- + \bar{v}_e + Q$$
$$L_e: \quad 0 \to 0 + (+1) + (-1);$$

and
$$\gamma \to \mathrm{e}^- + \mathrm{e}^+ + Q_\gamma$$
$$L_e: \quad 0 \to (+1) + (-1);$$

and
$$^{64}_{29}\mathrm{Cu} \to {}^{64}_{28}\mathrm{Ni} + \mathrm{e}^+ + v_e + Q^+$$
$$L_e: \quad 0 \to 0 + (-1) + (+1).$$

Smilarly for the muon lepton number,
$$\pi^- \to \mu^- + \bar{v}_\mu + Q_\pi$$
$$L_\mu: \quad 0 \to (+1) + (-1);$$

and for both, $\quad \mu^+ \to \mathrm{e}^+ + v_e + \bar{v}_\mu + Q_\mu$
$$L_e: \quad 0 \to (-1) + (+1) + 0$$
$$L_\mu: \quad (-1) \to 0 + 0 + (-1).$$

The muon has been the subject of some quite exceptionally accurate measurements of the ratio of its magnetic moment to its intrinsic spin. The aim is to test the predictions of quantum electrodynamics, which have been found to be 'perfect' for electrons within very small limits of error in studies of the energies and intensities of spectral lines for atomic hydrogen.

It is common to express the so-called gyromagnetic ratio as

$$g = \frac{\mu}{I} = \frac{\text{magnetic moment in units of } e\hbar/2m_\mu}{\text{intrinsic spin angular momentum in units of } \hbar},$$

and for both electrons and muons g is very close to 2. (The unit $e\hbar/2m_\mu$ is the magnetic moment of a particle of charge e and mass m_μ moving in a circle with angular momentum $\hbar$, neglecting any effects due to its intrinsic spin and magnetic moment.) An experiment at CERN, Geneva, measured $(g-2)/2$ for the muon and obtained 0.001162 (± 5 in the last place). The theoretical value is 0.001165, and this remarkable agreement is viewed as strong support for the validity of quantum electrodynamics. The absence of an 'anomalous' magnetic moment indicates that, like the electron, the muon is subject to weak and electromagnetic interactions (and gravity) but not to the strong interaction. It also implies that muons, like electrons (and, presumably, tauons) are structureless and can be considered to be points with their surrounding field quanta of photons and W's and Z's.

Hadrons

The term 'hadron' refers to all types of particles that interact with each other through the strong interaction. These particles are subdivided into baryons and mesons: the baryons act as 'sources' for the mesons, which are the nuclear field quanta. So far we have encountered the proton and neutron and their antiparticles, and the pi-meson or pion, with a few references to kaons.

Pions were first observed (Lattes, Occhialini, and Powell 1947) in nuclear emulsions exposed for several months to cosmic rays on a high mountain. They exhibited a decay into muons of a definite energy, indicating a two-body decay, now known to be

$$\pi^- \to \mu^- + \bar{\nu}_\mu + 33.9 \text{ MeV},$$

and

$$\pi^+ \to \mu^+ + \nu_\mu + 33.9 \text{ MeV}.$$

In the following year these pions were produced artificially by Lattes and Gardner (1948) by means of 380 MeV alpha-particles (from the Berkeley synchrocyclotron) impinging on carbon and other targets, and were again detected by nuclear emulsions.

Oppenheimer (1947) suggested that a neutral pi-meson, postulated by several workers as consistent with Yukawa's theory, would decay with a very short mean life into two gamma photons. In 1950 such energetic photons, of approximately 70 MeV, were observed from targets bombarded by 180–350 MeV protons and also from the interactions of 330 MeV gamma-rays derived from the Berkeley electron synchrotron. In 1951 the neutral-pion (π^0) rest mass was accurately deduced from the study of negative pion absorption in hydrogen which is either 'radiative absorption',

$$\pi^- + p \rightarrow n + \gamma + Q(E_\gamma \simeq 131 \text{ MeV}),$$

or 'mesic absorption'

$$\pi^- + p \rightarrow n + \pi^0 + Q$$
$$\quad\quad\quad\quad\quad \downarrow 2\gamma(E_\gamma \simeq 55 - 85 \text{ MeV}).$$

The 30 MeV spread of gamma energies arises from the Doppler shift due to the velocity of the decaying π^0, towards or away from the observer and leads to a value of the mass difference of the pions $m(\pi^-) - m(\pi^0)$ of 4.6 MeV/c^2. Since $m(\pi^-) = 139.6$ MeV/c^2, we have $m(\pi^0) = 135.0$ MeV/c^2. The decay into two gamma-photons, and not three, also allows a quantum-mechanical argument for zero spin of the π^0.

The identification of a 'new' particle requires at least one parameter that is different from that of other particles. For many new particles this was simply the mass, although caution is needed in comparing, for example, the masses of charged and uncharged pions or even of the proton and neutron. These same particles also warn against labelling particles as 'new' or 'different' on the basis of different observed mean lifetimes. On the other hand, differences of spin J invariably indicate different particles, and so do large differences of mass. The modes of decay are so constrained by conservation laws that they can be crucial in the identification of new particles.

In the year of the pion, Rochester and Butler (1947) observed a pair of tracks forming a V in a cloud chamber exposed to cosmic rays, and deduced that they arose from the decay of a neutral V-particle (now known as the K^0). Since the tracks appeared to be two pions they deduced

$$K^0 \rightarrow \pi^+ + \pi^- + Q,$$

with the mass of the K^0 about 800 m_e (later found to be about 975 m_e). In 1949, using nuclear emulsions, co-workers of Powell at Bristol (Brown *et al.* 1949) identified the K^+ in a decay to three pions,

$$K^+ \rightarrow \pi^+ + \pi^+ + \pi^- + Q,$$

which allowed a fairly accurate determination of the K^+ mass (966 m_e).

In 1951 a Manchester group discovered the first hyperon (i.e. baryon heavier than the neutron) by another V-shaped track, this time identifying the decay particles as a proton and a pion. The neutral parent is now known as a Λ^0 (lambda zero) of mass 1115.4 MeV/c^2, and the decay is

$$\Lambda^0 \rightarrow p + \pi^- + Q.$$

By 1953 many more particles had been discovered in cosmic-ray events notably the Σ (sigma) and Ξ (xi) hyperons, and about the same time developments in accelerators allowed more careful determinations of the particle masses, spins, lifetimes, and decay modes. A list is provided in Table 8.1, together with details of other properties, such as parity, isospin, and hypercharge (or strangeness) that are introduced on pp. 152 to 155.

The last of these baryons to be discovered was the Ω^- in 1964 (see p. 157 ff). The τ lepton was discovered in 1975. Since 1974 a series of so-called 'charmed' particles have been found, and also 'bottom' particles. Their partners, 'top' particles, are confidently expected to be discovered (see pp. 171 ff).

Particle resonances

Most of the hadrons mentioned above have mean lifetimes between 10^{-8} s and 10^{-10} s [exceptions are the two electromagnetic decays of π^0 and Σ^0, and those of the η_0 (3×10^{-19} s),

$$\pi^0 \rightarrow 2\gamma \quad (9 \times 10^{-17} \text{ s}) \quad \text{and} \quad \Sigma^0 \rightarrow \Lambda^0 + \gamma \quad (10^{-14} \text{ s}),$$

and, of course the proton (stable, $> 10^{30}$ y) and neutron (10^3 s)]. Lifetimes of 10^{-10} s are long enough to allow the hadrons to be formed in nuclear collisions, move away from their point of origin,

TABLE 8.1

Type	Symbol and charge	Mass (MeV/c^2)	Mean life (s)	Principal decays	Spin and parity	Isospin	Hypercharge
	γ	0	∞		1^-	0	0
Leptons	ν_e	0?	∞		$\frac{1}{2}$		0
	$e^\pm$	0.511	∞		$\frac{1}{2}$		0
	ν_μ	0?	∞		$\frac{1}{2}$		0
	$\mu^\pm$	105.7	2.2×10^{-6}	$e\nu_e\nu_\mu$	$\frac{1}{2}$		0
	ν_τ	0?	∞		$\frac{1}{2}$		0
	$\tau^\pm$	1784	$\sim 10^{-13}$ s	$\mu\nu_\mu\nu_\tau$	$\frac{1}{2}$		0
Mesons	$\pi^\pm$	139.6	2.6×10^{-8}	$\mu\nu$	0^-	1	0
	π^0	135.0	9×10^{-17}	$\gamma\gamma$	0^-	1	0
	$K^\pm$	493.8	1.2×10^{-8}	$\mu\nu, \pi\pi$	0^-	$\frac{1}{2}$	± 1
	K^0	497.8		50 per cent K_S^0 50 per cent K_L^0	0^-	$\frac{1}{2}$	± 1
	K_S^0	497.8	8.6×10^{-11}	$\pi\pi$	0^-	$\frac{1}{2}$	± 1
	K_L^0	497.8	5.3×10^{-8}	$\pi\pi\pi, \pi\mu\nu$ $\pi e\nu$	0^-	$\frac{1}{2}$	± 1
	η^0	548.8	3×10^{-19}	$\gamma\gamma, \pi\pi\pi$	0^-	0	0
Baryons	p	938.3	∞?		$\frac{1}{2}^+$	$\frac{1}{2}$	1
	n	939.6	960	$pe\nu$	$\frac{1}{2}^+$	$\frac{1}{2}$	1
	Λ^0	1115	2.5×10^{-10}	$p\pi, n\pi$	$\frac{1}{2}^+$	0	0
	Σ^+	1189	8×10^{-11}	$p\pi, n\pi$	$\frac{1}{2}^+$	1	0
	Σ^0	1193	$\sim 10^{-14}$	$\Lambda\gamma$	$\frac{1}{2}^+$	1	0
	Σ^-	1197	1.6×10^{-10}	$n\pi$	$\frac{1}{2}^+$	1	0
	Ξ^0	1315	3×10^{-10}	$\Lambda\pi$	$\frac{1}{2}^+$	$\frac{1}{2}$	-1
	Ξ^-	1321	1.6×10^{-10}	$\Lambda\pi$	$\frac{1}{2}^+$	$\frac{1}{2}$	-1
	Ω^-	1672	1.3×10^{-10}	$\Xi\pi, \Lambda K$	$\frac{3}{2}^+$	0	-2

and subsequently decay. The characteristics of the corresponding tracks are fairly obvious, making allowance for 'gaps' where a neutral particle leaves no track. A whole new range of 'particles' has now been discovered with very much shorter lifetimes. They require quite different techniques for study, the first of which, 'resonance scattering', was investigated as early as 1952 (see p. 150), although its significance was not realized until higher energies allowed 'resonance production' in 1961. The latter is perhaps easier to follow than the former.

Consider the reaction A(a, b)B. It is clear that the definite energies, kinetic and rest-mass, before the reaction in conjunction with linear-momentum conservation require the two final particles (assumed to be stable) to have definite kinetic energies. On the other hand, a reaction between two initial particles leading to three stable particles A(a, bc)B allows the kinetic energy and linear momentum to be shared in an infinite variety of ways between b, c, and B, provided the total energy and linear momentum are conserved. If now we consider the case where two of the three final particles live as 'one particle' $(b + B)$ for a finite mean lifetime τ then the resultant 'two-body' decay to b and B will have an energy uncertainty δE associated with it such that $\delta E . \tau \simeq \hbar$. The result is a tendency towards the definite final energies $T_{(b + B)}$ and T_c of the two-body final state, going over to the energy spread of the three-body final state when τ, the mean life of the so-called 'resonant particle', happens to be exceedingly short. Since $\hbar = 6.582 \times 10^{-22}$ MeV s, energy spreads of $T_{(b + B)}$ and hence T_c as large as 65 MeV can be obtained for mean lives of $\tau \simeq 10^{-23}$ s, and many particles with such mean lives have been observed.

A most effective way of appreciating the analysis of these energy spreads is by means of a Dalitz plot. An example is shown in Fig. 8.3(a) for the twin reactions

$$K^- + p \rightarrow (\Lambda^0 + \pi^+) + \pi^-$$

and

$$K^- + p \rightarrow (\Lambda^0 + \pi^-) + \pi^+ \quad (T_{K^-} = 758 \text{ MeV}).$$

The resonant particle $(\Lambda^0 + \pi^+)$, now known as the $\Sigma^+(1385)$, is formed with a mean mass of 1385 MeV/c^2, and if it were long-lived the kinetic energy of the π^- in the centre-of-mass system would be about 285 MeV. Similarly, for the $\Sigma^-(1385)$ reaction, T_{π^+} would be 285 MeV. For each event observed in a hydrogen bubble chamber the value of T_{π^-} is plotted against T_{π^+} and it is seen that both T_{π^-} and T_{π^+} are spread about the value 285 MeV. If, on the other hand, the reaction had produced three particles (i.e. the 'lifetime' of the $\Sigma(1385)$ had been exceptionally short, say $< 10^{-30}$ s), the coordinates of the plotted points would have been uniformly spread within the envelope of the curve shown, which is set by the joint conservation of energy and linear momentum. For each event a mass of the $\Sigma(1385)$ can be calculated and the result is given in Fig. 8.3(b). Later work has shown that the mass is nearer 1382 MeV/c^2 with a mean life of $\sim 1.8 \times 10^{-23}$ s corresponding to $\Delta E \sim 36$ MeV.

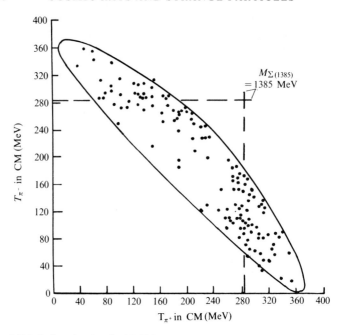

Fig. 8.3(a) Dalitz plot for the $\Sigma(1385)$ resonances produced by 758 MeV negative kaons on protons. (See Fig. 8.3(b).)

Resonance scattering of pions by protons was first recognized by Fermi and co-workers (Anderson *et al.* 1952) and is indicated by an enhanced probability of scattering near a particular energy of the incident pions (Fig. 8.4). Similar enhanced scattering had been observed for neutrons on nuclei and interpreted in terms of a compound nucleus of mean life τ given by $\tau\Gamma = \hbar$, where Γ is the width of the resonance, being (in the simplest case) the energy width of the cross-section curve at half its peak value. For the π^+p system there is clearly a resonant state (or particle) of mass 1236 MeV/c^2 with $\Gamma \simeq 120$ MeV/c^2, corresponding to a mean life of 5.5×10^{-24} s. This is known as the $\Delta(1236)$, and from its presence also in π^-p scattering it is clear that it can have two units of positive charge, Δ^{++}, or zero charge, Δ^0. Later studies showed that Δ^+ and Δ^- also exist (see Fig. 8.7). Figure 8.4 indicates several more resonant particles. A continuing study of the scattering of different particles (π, K, etc.) over wide ranges of energy has led to the discovery of a great number of resonances. They can be broadly classified into those that decay into

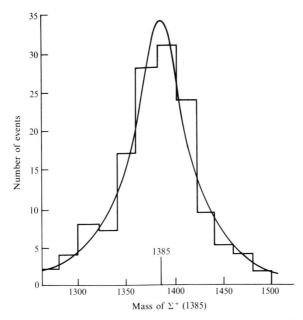

Fig. 8.3(b) Mass spectrum of the Σ(1385) resonances. (Adapted from M. H. Alston *et al.* (1960.))

baryons and mesons (baryon resonances) and those that decay into mesons only (meson resonances). Notice that some resonances can be observed only in resonance-production reactions and not in resonance scattering. A good example is the Σ(1385) discussed above since, in order to be seen in resonance scattering it would require a beam of pions on a Λ^0 target, or of Λ^0 on a pion target, and each is manifestly impracticable.

In 1974 the first of a new generation of particles was discovered (the J/ψ, see p. 166), many of which had been conjectured or predicted by theoreticians. In order to convey the sense of an unfolding story, we shall not present the fullness of the present picture until we have followed the attempts to classify the particles known up to about 1970—attempts that were remarkably successful in preparing the ground for new discoveries.

Classification of hadrons

Hitherto we have recognized only four identifiable characteristics of hadrons: their mass, electric charge Q, spin J and lifetime. For leptons

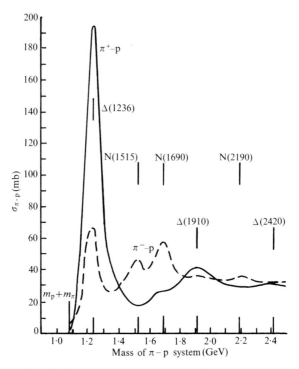

Fig. 8.4. Resonance scattering of π^+ and π^- by protons.

we introduced the lepton numbers L_e, L_μ, L_τ, and experience has shown that a baryon number B is definable, so that it too is conserved in all types of reactions. Positive values apply to baryons and negative values to antibaryons. $B = 1$ for p and n, and $B = -1$ for $\bar{p}$ and $\bar{n}$; and similarly for the hyperons (Λ, Σ, Ξ, Ω) and antihyperons. No particle with $|B| > 1$ has been found in high-energy production processes, although atoms other than ordinary hydrogen have nuclei with $B = A > 1$. Mesons and leptons have $B = 0$.

The nuclear similarity of the proton and neutron, quite apart from their similar masses and identical spin ($\frac{1}{2}$), has been deduced from many distinct observations. These range from simple facts such as the existence of approximately equal numbers of stable nuclei with Z even $- N$ odd and Z odd $- N$ even, and the near equality of the separation energies S_p and S_n for protons and neutrons from comparable nuclei, through to the detailed analysis of high-energy scattering of protons and neutrons. It seems that if we could 'switch off' the electromagnetic interaction the proton and neutron would be indistinguishable. In other words, the neutron and proton appear to be two 'charge states' of the same nuclear particle. As early as 1932 Heisenberg recognized some of the advantages of a formalism parallel to that of the ordinary intrinsic spin of $\frac{1}{2}$ which can be either 'up' or 'down'. Wigner (1937) developed this idea and assigned an 'isospin' of $T = \frac{1}{2}$ for the nucleon, with the values $+\frac{1}{2}$ for the proton and $-\frac{1}{2}$ for the neutron. More accurately, $T_3 = +\frac{1}{2}$ for the proton, and $T_3 = -\frac{1}{2}$ for the neutron, where T_3 is the third component of isospin—in exact analogy to the 'z-component' of ordinary spin, S_z, where z is a direction impressed by the observation, e.g. the normal to the scattering plane, or the direction of an applied electric or magnetic field. (See p. 192.) Isospin space has no relation to ordinary space other than the analogous formalism.

The proton and neutron form an isospin doublet, and other particles fit into other 'multiplets': singlets ($T = 0$), triplets ($T = 1$), and quadruplets ($T = \frac{3}{2}$) the 'multiplicity' being $2T_3 + 1$. Closer inspection reveals that the electric charge Q is given by the formula

$$Q = T_3 + \tfrac{1}{2}B$$

for nucleons and for pions, provided that $T = 1$ for the latter (thus T_3 of $\pi^- = -1$, of $\pi^0 = 0$, and of $\pi^+ = +1$, since $B = 0$ for pions). However, for some other particles this formula appears not to be valid. Thus for the Λ^0 which decays to $p + \pi^-$ or to $n + \pi^0$, it is evident that $B = 1$, but since the Λ^0 is equally obviously an isospin singlet with $T = T_3 = 0$, the formula suggests $Q = \frac{1}{2}$! This was one of a number of problems encountered in studies of hyperons and kaons.

The probability of Λ^0 production by π^- bombardment of protons is quite high (e.g. 10 mb cross-sections are found), and it is clear that the strong interaction is involved, i.e. production is in times of $\sim 10^{-23}$ s. But the decays with mean lifetimes $\sim 10^{-10}$ s are some 10^{13} times as slow, i.e. decay is by the weak interaction, and this is very strange. As Wilkinson remarks, 'So long a lifetime appears to be a suspension of the normal processes of nature and is quantitatively even worse than if Cleopatra had fallen off her barge in 30 BC. and had taken until last week to splash into the Nile, which would have been too long by a factor of only 10^{11}.' A clue to the solution was noted by Nambu *et al.* (1951) and Pais (1952), namely, that the 'strange' particles, as they came to be called, were produced in association with each other, although of course they decay in isolation, e.g.

$$\pi^- + p \to \Lambda^0 + K^0.$$

Gell-Mann (1952) and, independently, Nakano and Nishijima (1952) accounted for these and other phenomena by the introduction of a new quantum number initially designated 'strangeness' S. It is part of the fun of physicists, especially in this field, to choose amusing nomenclature. Other related quantum numbers are now well established, namely 'charm', 'topness' (or 'truth'), and 'bottomness' (or 'beauty')—see pp. 166, 171. For strong and electromagnetic interactions strangeness is conserved ($\Delta S = 0$), but for weak interactions it is not conserved and indeed $|\Delta S| = 1$. Systematic assignments starting with the arbitrary values $S = 0$ (appropriately enough) for proton and opposite values for $\Lambda^0 (S = -1)$ and K^0 ($S = +1$) led eventually to a coherent scheme, thus:

production
$$\pi^- + p \to \Lambda^0 + K^0 \quad (\Delta S = 0)$$

decay (e.g.)
$$\Lambda^0 \to p + \pi^- \quad (\Delta S = +1)$$

and (e.g.)
$$K^0 \to \pi^+ + \pi^- \quad (\Delta S = -1).$$

A new formula is now needed to link electric charge and isospin, namely,

$$Q = T_3 + \tfrac{1}{2}B + \tfrac{1}{2}S,$$

leading to the alternative nomenclature $Q = T_3 + \tfrac{1}{2}Y$, where $Y = B + S$ and is termed 'hypercharge'.

An interesting and typical consequence of the development of the Gell-Mann strangeness scheme was the prediction of a 'new' particle. Σ^+ and Σ^- had been observed, but not Σ^0, suggesting $T = \tfrac{1}{2}$ for the Σ. Decays such as $\Sigma^+ \to \pi^+ + n$ indicate $B = 1$ and $|S| = 1$, and therefore suggest $Q = +\tfrac{1}{2}$ or $\tfrac{3}{2}$, which was not observed. A solution is found if Σ^0 exists, since then $T = 1$ and $S = -1$ satisfies all requirements with $T_3 = +1, 0, -1$ for Σ^+, Σ^0, and Σ^-, respectively. The experimenters duly obliged in 1957 by discovering the Σ^0 in analyses of $\pi-$ reactions in a hydrogen bubble chamber. Since $\Sigma^0 \to \Lambda^0 + \gamma$, $\Delta S = 0$, and the reaction is therefore relatively quick by means of the electromagnetic interaction ($\tau \sim 10^{-14}$ s). It was rather obviously difficult to observe because of the neutral particles, the full process being

$$\pi^- + p \to \Sigma^0 + K^0$$
$$\big\lfloor\!\!\longrightarrow \Lambda^0 + \gamma$$

Our identifying characteristics for hadrons are now six; mass, spin, baryon number, isospin (T_3), hypercharge and lifetime. One more is available from quantum mechanics, namely, the intrinsic parity of the wavefunction representing the particle (see p. 98). Only changes of parity can be deduced from observations, coupled with quantum-mechanical considerations, and we choose the intrinsic parity of the proton to be positive. It is common to combine spin J and parity π in a single symbol J^π. Thus for a proton $J^\pi = \frac{1}{2}^+$, and for a pion 0^-. For many years it was assumed that the total parity of a system is conserved for all types of interactions, but Lee and Yang (1956) realized that non-conservation of parity in the weak interaction could explain some curious features of the weak interaction decays of kaons, and they proposed experimental tests using the beta-emission of ^{60}Co (see p. 160). These experiments quickly proved the non-conservation of parity under the conditions of a weak interaction, but it is still believed to be conserved in the strong and electromagnetic interactions. One conclusion from the experiments is that the neutrino spin and linear-momentum vectors are oppositely directed, as for a left-handed 'screw. This led Heisenberg to quip that 'God must be a weak left-hander'.

A preliminary classification is now possible. For baryons $B = 1$ we have the 'families' shown in Table 8.2 which include not only the nucleons and hyperons but also the particle resonances. For mesons ($B = 0$) there are three 'families' characterized as π with $Y = 0$, $T = 1$; η with $Y = 0$, $T = 0$; and the rather complicated K with $Y = +1$, $T = \frac{1}{2}$. We shall not pursue the fascinating story of the kaons because of its

TABLE 8.2

Baryon family		Hypercharge (Y)	Isospin (T)
Nucleon	N	1	$\frac{1}{2}$
Lambda	Λ	0	0
Sigma	Σ	0	1
Xi	Ξ	-1	$\frac{1}{2}$
Omega	Ω	-2	0
Delta	Δ	1	$\frac{3}{2}$

numerous facets and the range of new quantum-mechanical concepts involved.

The introduction of isospin enabled particles to be grouped together in a unifying way, the small differences in masses, for example between m_p and m_n, being accounted for by the 'splitting' produced by the electromagnetic interaction. It was found to be profitable to try to relate certain isospin multiplets of much greater mass differences in larger groupings, with the thought that other interactions, possibly new ones, can account for the 'mass splitting'. The mathematical formalism suitable for spin and isospin multiplets is termed SU(2), meaning 'special unitary group for arrays of size 2×2'. In Appendix D we develop some of this isospin formalism in order to introduce the powerful formalisms used in a more complete classification of particles, in particular SU(3). In 1960 the extension of SU(2) to SU(3) was tried with remarkable success by Gell-Mann (1961), and independently by Ne'eman (1961), first for groups (or so-called 'supermultiplets') of eight baryons, and of eight mesons, and later for a group of ten baryons, and for singlets.

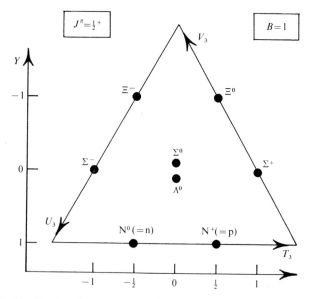

Fig. 8.5. Classification of baryons of $J^\pi = \frac{1}{2}^+$ suggested by unitary symmetry SU(3). This octet includes the proton and neutron. (It is drawn in relation to a triangle rather than a hexagon to emphasize the threefold symmetry.)

In Fig. 8.5 based on SU(3) theory the particles with $J^\pi = \frac{1}{2}^+$ are displayed as functions of their hypercharge Y and isospin T_3; they form an octet. The predicted relation between the masses works well,

$$2M_N + 2M_\Xi = 3M_\Lambda + M_\Sigma,$$

and it is evident that the relation could have been displayed in a closely similar manner by successive rotations through 120°, so that the lines of symmetry labelled U and V successively take the place of the isospin ordinate T_3. The interpretation of so-called U-spin and V-spin is given in Appendix D.

For the familiar group of mesons with $J^\pi = 0^-$ the pattern is shown in Fig. 8.6, and is also an octet, but in 1960 the η^0 had not been observed. It was therefore sought by the experimentalists, its mass having been estimated as 567 MeV/c^2 from the predicted relation

$$2M_K^2 + 2M_{\bar{K}}^2 = 3M_\eta^2 + M_\pi^2.$$

In 1961, the η^0 was found, with a mass of 549 MeV/c^2 and a mean lifetime of about 3×10^{-19} s—sufficiently close to the predictions to encourage the theoreticians.

A more striking prediction of SU(3) theory followed from the baryon decuplet (Fig. 8.7) with $J^\pi = \frac{3}{2}^+$. In 1961 the Ω^- had not been

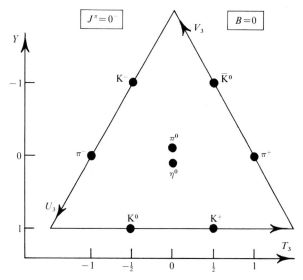

Fig. 8.6. Octet of mesons of $J^\pi = 0^-$, which led to the prediction of the η^0.

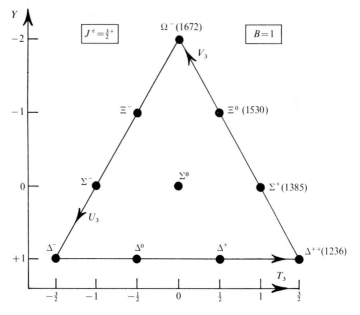

Fig. 8.7. Decuplet of baryons of $J^\pi = \frac{3}{2}^+$ which led to the prediction of the Ω^-.

found in spite of its 'long' life ($\sim 10^{-10}$ s), although its properties could be predicted with some confidence. Its expected mass of 1675 MeV/c^2 followed from

$$M_\Omega - M_\Xi = M_\Xi - M_\Sigma = M_\Sigma - M_\Lambda,$$

its electric charge would be -1 from $Q = T_3 + \frac{1}{2} Y$, and its hypercharge of -2 (strangeness -3) together with its mass implied that its decay would be to $\Xi^0 + \pi^-$, $\Xi^- + \pi^0$, or $\Lambda^0 + K^-$. All of these decays are by weak interactions ($|\Delta S| = 1$) and are therefore slow, whereas all the other members of this decuplet are particle resonances with $\Delta S = 0$, i.e. with fast decays ($\sim 10^{-23}$ s). After a great deal of searching in hydrogen-bubble-chamber pictures of interactions of K^- ($Y = -1$) mesons, the Ω^- was discovered in 1964 (Fig. 5.6) with a mass of 1672 MeV/c^2. The event was hailed as a triumph for the SU(3) theory of classification, with far-reaching consequences.

Problems

8.1. In Anderson's first cloud-chamber picture of the positron (Fig. 8.2(a)) the radius of the track of the incident particle is 140 mm, and after passing through a 6.0 mm lead plate the track radius is 51 mm. The applied magnetic flux density B was 1.50 T, and the plane of each track was approximately perpendicular to B. (a) Calculate the initial and final energies of the positron and compare its mean rate of energy loss with that for an electron, as shown in Fig. 5.1(b). What can be deduced from the droplet densities of the two tracks? (b) Show that the incident particle could not have been a proton. The semi-empirical range–energy relation $R(m) = \{ T(\text{MeV})/9.3 \}^{1.8}$ for protons in air, is valid from a few MeV to 200 MeV. Is this relevant to the problem?

8.2. A muon of rest mass 105.7 MeV/c^2, moving in a cloud chamber at right angles to a magnetic field of flux density 1.00 T, passes through a metal plate. The radii of curvature of its coplanar tracks before and after the plate are 0.50 m and 0.25 m respectively. Find the energy (in MeV) lost in the plate. Compare the answers obtained by the use of relativistic and non-relativistic formulae.

8.3. Show that a photon cannot spontaneously produce an electron–positron pair in the absence of another particle.

8.4. Calculate the energies of the muon and muon–neutrino produced in the decay of a stationary pion.

$$(M_{\pi^+} = 139.6 \text{ MeV}/c^2, \ M_{\mu^+} = 105.7 \text{ MeV}/c^2.)$$

9. More particles, and cosmic consequences

Conservation laws and symmetry

The conservation 'laws' of physics were for some 300 years restricted to those of mass, linear momentum, and angular momentum. In 1905 Einstein's theory of relativity developed the concept of mass–energy as the conserved quantity rather than mass alone. The conservation of electric charge, in integral multiples of the electronic charge, came to be recognized, without being formally proposed, and for the first half of the twentieth century these four conservation laws formed the basis for the whole structure of physical laws or quantifiable relations. Just before the success of Fermi's neutrino theory there had been some doubts expressed, notably by Bohr, but all-in-all the four laws remained intact.

It seemed so obvious that the mirror image of a physical system would behave in fundamentally the same sort of way as the original system that the concept of parity conservation in quantum-mechanical treatments was assumed almost without question. It was the τ–θ puzzle in particle physics that led to the suggestion that parity may not be conserved in the weak interaction. The τ-meson is not to be confused with the τ-lepton or tauon—the nomenclature τ-meson and θ-meson is now abandoned since these particles are realized to be the same as the K-meson or kaon. In the early 1950s the τ and θ were found to have identical masses and lifetimes but different decays, the τ to three pions and the θ to two pions (see p. 147). It was deduced from parity conservation that the τ and the θ must therefore have opposite intrinsic parity and be different particles. But this was such an awkward result that Lee and Yang (1956) proposed tests of parity conservation.

The first evidence of parity non-conservation in the weak interaction came from experiments by Wu et al. (1957) on the β^--decay of polarized ^{60}Co nuclei (Fig. 9.1). The nuclear spins are aligned in the opposite direction to the applied magnetic field produced schematically by the coil carrying current I, the polarization being enhanced

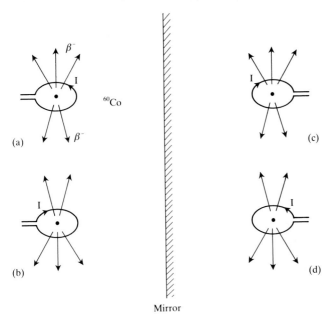

Fig. 9.1. Demonstration of parity non-conservation in the β^--decay of polarized ^{60}Co nuclei.

by very low temperatures (~ 0.01 K). It is found (Fig. 9.1(a)) that more β's are directed parallel to the magnetic field than anti-parallel. When the current, and therefore the polarization, are reversed the asymmetry is reversed (Fig. 9.1(b)). If we had observed these experiments in a mirror we would have seen the results shown in Fig. 9.1(c) and (d). Clearly (c) contradicts (b), and (d) contradicts (a), hence reflection symmetry is not observed, i.e. parity is not conserved. If parity were conserved there would have to be no asymmetry of the β's. The way was therefore open for the τ-meson to be identified with the θ-meson, thereby simplifying the classification of particles. Notice that there is not total non-conservation, i.e. parity is not always changed, for that would reintroduce the τ–θ problem. Sometimes the kaon decays into two pions, and sometimes into three pions.

Parity is an example of a 'discrete symmetry' which leads to 'multiplicative' conservation laws. Thus the parity of a system, $+1$ or -1, is the product of the parities of its component parts, and clearly imposing reflection twice over produces the original, corresponding to $(-1) \times (-1) = +1$. The non-conservation of parity in the weak interaction came as something of a shock, and led to more searching investigation of the underlying assumptions in other conservation laws.

It had been known for some time that conservation of mass–energy followed as a deducible necessity from the observation that the behaviour of a given physical system is the same today as it was yesterday, i.e. that it is independent of translation in time. Correspondingly, independence of a translation in space implies conservation of linear momentum (Appendix E), and independence of rotation through a given angle in space implies conservation of angular momentum. These laws arise from 'continuous symmetries' which produce 'additive' conservation laws, since the total energy (for example) of a system is obtained by adding up the energies of the several parts. Conservation of electric charge cannot be related to such commonsense 'symmetry operations', but is fully understood in terms of so-called gauge invariance (i.e. electric and magnetic fields are independent of shifts in the zero values of the scalar and vector electromagnetic potentials V and A).

What can be said about the conservation of lepton numbers L_e, L_μ and L_τ, and the baryon number B? So far we have no known symmetry operations that imply these three conservation laws. It is a great challenge to try to identify those 'obvious' unconscious assumptions about the nature of the physical world that imply the conservation of L_e, L_μ, L_τ and B. On the other hand, some theories that aim to unify the electroweak and the strong interactions require the non-conservation of the three Ls and B, and there are important experiments in progress looking for the decay of the proton (see p. 177).

Other symmetry operations that have been already identified include 'time-reversal', and 'charge-conjugation'. If a cine-film of a physical event can be run backwards and the time-reversed events follow known physical laws it is clearly impossible, in principle, for an independent observer to know whether the film is running forwards or backwards. This is a highly simplified presentation of what is known as the conservation of 'time-parity' T. Charge-conjugation is the act of symmetry operation in which every particle in a system is replaced by its antiparticle. If the anti-system, or anti-matter counterpart, exhibits the same physical phenomena then 'charge-parity' C is conserved. In fact, C (like space-parity P) is not conserved in the weak interaction (a deduction from the right- and left-handedness re-

spectively of positrons and electrons from the decay of μ^+ and μ^-).
Both C and T are multiplicative quantum numbers, taking values $+1$
or -1, which are conserved for the electromagnetic and strong
interactions.

The combined symmetry operation in which the anti-matter mirror
image of a system is run in reverse allows a test of CPT invariance,
first postulated by Schwinger (1953), one predicted consequence of
which is that particles and their antiparticles have identical masses
and lifetimes. All the evidence, some of high precision, supports the
conservation of CPT. Furthermore, we do not know how to construct
theories that are not invariant under CPT. On the other hand, there is
good reason to believe, from observations of decay modes of neutral
kaons (Christenson *et al.* 1964) that conservation of the combined
parities CP does not hold in weak interactions, and if CPT is
conserved then T cannot be conserved in weak interactions. The
present position is summarized in Table 9.1, where a tick implies
conservation and a cross non-conservation.

TABLE 9.1

	Strong	Electro-magnetic	Weak
CPT	✓	✓	✓
P	✓	✓	×
C	✓	✓	×
CP or T	✓	✓	Slightly ×

Quarks and the discovery of charm

The naming by Rutherford of the alpha and beta rays or particles was
doubtless based on a justifiable belief in his priority of identification of
these two species of radiation. It would be nice to know whether or
not the implied finality of the name omega, used in the 1961
prediction of the last 'long-lived' (1.3×10^{-10} s) hyperon to be
discovered, was deliberate. No other such hyperon had at that time
been predicted, but within a few years new predictions matched by
new discoveries led to major advances, not just in the numbers of
particles but in fundamental understanding. The large numbers of
particles and particle resonances already identified by the end of the
1950s strongly suggested that there must be some other more
fundamental or more elementary units of which all these particles are

composed. Out of the many suggestions, the quark theory came to be favoured. The first tentative related suggestion was made by Sakata (1956), namely that all the particles were made from combinations of p, n, and Λ, but this was soon shown to be untenable.

The next major step forward was by Gell-Mann (1961) and independently by Ne'eman (1961), by involving SU(3) group theory to classify the known particles (see pp. 156 ff and Appendix D). It was the essential ingredient of 'threeness' that led Gell-Mann to the name 'quark', as found in a line taken from James Joyce's *Finnegans Wake*,

'Three quarks for Muster Mark'

—this implies that 'quark' should rhyme with Mark and the bark, lark, dark, and Park of the following lines, and not, as is generally believed and accepted by the practitioners, rhyme with 'corks' and 'quarts'. The three quarks for Muster Mark are usually interpreted as his children and not his drinking capacity!

The theme of 'threeness' soon came to be taken more seriously and it was shown by Gell-Mann (1964) and independently by Zweig (1964) that all the baryons known at that time could be constructed from three quarks, and all mesons from quark–antiquark pairs. These three quarks are now referred to as u, d, s, with 'up', 'down', and 'strange' so-called 'flavours', and they have the properties displayed in Table 9.2.

TABLE 9.2

Properties of the first three quarks

Quark	B	T	T_3	Y	Q	S	Mass (MeV/c^2)
u	$\frac{1}{3}$	$\frac{1}{2}$	$\frac{1}{2}$	$\frac{1}{3}$	$\frac{2}{3}$	0	336
d	$\frac{1}{3}$	$\frac{1}{2}$	$-\frac{1}{2}$	$\frac{1}{3}$	$-\frac{1}{3}$	0	336
s	$\frac{1}{3}$	0	0	$-\frac{2}{3}$	$-\frac{1}{3}$	-1	500–700

There are also three antiquarks with the same corresponding values for T and mass, but all other values of opposite sign to those for the quarks. The proton is made up of uud, the neutron udd, the Σ^+ is uus, the π^+ is u$\bar{\text{d}}$, the K^+ is u$\bar{\text{s}}$, the K^0 is d$\bar{\text{s}}$, the Ω^- is sss, and the Δ^{++} is uuu.

A striking property of quarks is their fractional electric charge, $\frac{1}{3}e$ or $\frac{2}{3}e$, which should be highly characteristic if it were possible to detect isolated quarks. Notice that if quarks were once formed, the chances of their meeting an antiquark with opposite charge with which they could annihilate would be very small. Their masses can be large, with compensating large binding energies. However, free quarks, which should be stable, have never been observed in spite of many searches for their high-energy tracks in suitable detectors, and also in electrostatic Millikan–oil-drop types of experiments. It is generally believed that attempts to knock out quarks from protons fail because of the rapid increase of the quark–quark interaction with separation distance such that (see p. 129) quark–antiquark pairs, e.g. pions, are preferentially produced from the energy supplied. The most direct experimental evidence for quarks is found in 'deep inelastic' scattering of GeV electrons by protons. The angular distributions with larger than expected numbers at large angles can be satisfactorily accounted for only if it is postulated that there are point-like charged objects present. This is reminiscent of the large-angle scattering of alpha-particles by heavy atoms that led Rutherford to postulate a 'point-like' positively charged nucleus (see p. 35). Comparison of the electron scattering results with those from experimentally difficult neutrino–proton scattering indicates that charges of $\frac{1}{3}e$ and $\frac{2}{3}e$ for these point-like objects agree with the data. Although initially termed 'partons', it is now accepted that these proton constituents are indeed quarks.

Table 9.2 shows only three quarks but already, in the same year as this highly successful proposal, Bjorken and Glashow (1964) suggested there should be a fourth. They pointed out that the leptons then known occurred in pairs, (e, v_e) and (μ, v_μ), and in weak interactions there is a lepton–hadron symmetry. Noting that leptons and quarks were both point-like, and of spin $\frac{1}{2}$, they indicated the possibility of a lepton–quark symmetry. The flavour of this fourth possible quark was given the name 'charm', and the corresponding pairs were (u, d) and (s, c).

Later in the 1960s it was concluded that the weak interaction required not only charged intermediate vector bosons, the W^+ and W^- (see pp. 127 ff) but also the neutral Z^0. Various 'neutral current' interactions (see p. 170) should then be possible, such as neutrino–proton and neutrino–electron elastic scattering (discovered in 1973) and also such decays as $K^0 \rightarrow \mu^+ + \mu^-$. When these decays were not observed it was suggested by Glashow et al. (1970) that yet another quantum number was needed, in close parallelism to the introduction of strangeness which explained why the 'fast' decay of certain (strange) particles was suppressed and the slow weak interaction took place instead. The new quantum number was identified with 'charm'.

The increase from three quarks (uds) to four (udsc) allows a new range of baryons and mesons to be predicted, using the SU group of next higher rank to SU(3), namely SU(4). Thus the (cc̄) meson would have a total charm number of zero, which in decay should reveal 'hidden' charm (or 'concealed' charm). And (cs̄), for example, with non-zero total charm is said to have 'revealed' charm (or 'naked' charm).

Once again theoretical speculations were rewarded by significant experimental discoveries. In late 1974, a neutral particle of mass $3095 \ \mathrm{MeV}/c^2$ was discovered at Brookhaven during 29 GeV proton bombardment of beryllium (Aubert *et al.* 1974). Independently, and almost simultaneously, the same particle was identified in electron–positron collisions in intersecting storage rings at Stanford where, within a few weeks, related particles were also found (Augustin *et al.* 1974). The lifetime of the so-called J/ψ particles, $\sim 10^{-20}$ s, is about 1000 times longer than those for typical resonances, corresponding to uncertainties in mass of less than $1 \ \mathrm{MeV}/c^2$. The J/ψ ($= \mathrm{c}\bar{\mathrm{c}}$) might at first be expected to decay by a mode similar to that of the ϕ (phi-meson, ss̄), which decays into a strange particle (K) and its strange antiparticle ($\bar{\mathrm{K}}$). But J/ψ cannot go to a charmed particle and its charmed antiparticle pair because the lightest such particles are too heavy (e.g. D^0 (1863) and $\bar{\mathrm{D}}^0$ (1863)) — hence the decay into $\mathrm{e}^+\mathrm{e}^-$ or $\mu^+\mu^-$. This indicated that J/ψ was indeed the predicted (cc̄) meson. The discovery of excited states, e.g. ψ' at $3684 \ \mathrm{MeV}/c^2$, with close parallels to the excited states of positronium ($\mathrm{e}^-\mathrm{e}^+$) led to the naming of J/ψ as 'charmonium'. The J/ψ has c, c̄ spins parallel, and the particle η_c with c, c̄ spins antiparallel has also been found, with the lower mass $2980 \ \mathrm{MeV}/c^2$. The effective mass of a c quark is in the range 1.5–$2 \ \mathrm{GeV}/c^2$.

The introduction of charm requires the two-dimensional relations of Figs 8.5–8.7 to be extended to three dimensions (Fig. 9.2). The (c, c̄) particle, i.e. the J/ψ, then appears at the centre, and the charmed particles and antiparticles are out of the Y, T_3 plane. Their masses have been predicted and agree well with experimental values, e.g. the excited state of the J/ψ, ψ'' (3772), decays to D^0 (1863) and $\bar{\mathrm{D}}^0$ (1863) which in turn decay respectively to $\mathrm{K}^-\pi^+$ and $\mathrm{K}^+\pi^-$ with lifetimes of

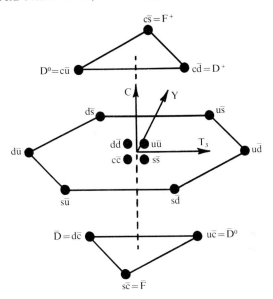

Fig. 9.2. Three dimensional display of mesons constructed from four quarks and four antiquarks, producing three new charmed mesons D^0, D^+, F^+ and their antiparticles, and the $(c, \bar{c})$ particle identified with the J/ψ particle for the case where c and $\bar{c}$ have parallel spins.

10^{-12} s. Other three-dimensional displays are needed to encompass the large numbers of new particles, for example the baryon decuplet triangle for Fig. 8.7 becomes the base with $C = 0$ of a tetrahedron with three more layers for $C = 1, 2$ and 3, containing successively 6, 3 and 1 new particles, the last being ccc. A fair number of these predicted particles have now been discovered.

The quarks are fermions, since they have a spin of $\frac{1}{2}$, and they are therefore expected to conform to the Pauli Exclusion Principle whereby no two fermions in a system can have the same quantum numbers. This immediately creates a problem for the quark theory of baryons since not only are there two u's (and a d) in the proton, and two d's (and a u) in the neutron, but there are no less than three s's in the Ω^-, three u's in the Δ^{++} and three d's in the Δ^- with the same quantum numbers. Rather than abandon the exclusion principle, Han and Nambu (1965) assigned an extra quantum number to each of the three quarks in a baryon, each taking one of three different values.

This is now referred to as the 'colour' of the quark, red or green or blue—not, of course, to be taken literally! Furthermore, the total colour of a particle is always 'white', either by including one of each colour for baryons, or a colour quark and the same colour antiquark as in mesons. These three colours form another SU(3) group (see Appendix D) which can be used to describe all possible arrangements of the three colours.

There are now several observations that confirm the need for the three colours. The ratio R, defined by

$$R = \frac{\sigma(e^+ e^- \rightarrow \text{hadrons})}{\sigma(e^+ e^- \rightarrow \mu^+ \mu^-)}$$

with σ the cross-section i.e. probability for the alternative processes, is predicted to be $\frac{2}{3}$ for no colour and 2 for the three-colour theory and is found by experiment to be about 2.5, for energies below the charmed-particle production threshold. Also the observed decay rate for $\pi^0 \rightarrow \gamma\gamma$ is nine times larger than the value from the no-colour theory, but in close agreement with the colour-theory prediction.

Three quarks in a baryon require an inter-quark force to hold them together and following the patterns for the other forces we expect there to be field bosons. These are termed 'gluons', possess unit spin, and are of two principal types, those that produce an exchange of colour between two quarks and those that do not. Eight different gluons are found to be required to bind the coloured quarks, and they possess eight different colour-carrying properties. The same gluons bind the quark–antiquark pairs in the mesons.

The formal theory treating such interactions is known as quantum chromodynamics (QCD) which has some analogies to QED, quantum electrodynamics, but there are also important differences. Thus a photon does not interact directly with another photon, but QCD indicates that gluons interact with gluons and may form detectable entities labelled 'glue-balls' with a total colour equivalent to 'white'. Confirmatory evidence of the presence of gluons in protons is provided by the analysis of electron and neutrino deep inelastic scattering (see p. 165). It is found that only half of the momentum of the parent proton is carried away by the three quarks. The other half is carried by constituents that do not interact with the electromagnetic or weak probes, i.e. by gluons that experience only the strong (interquark) interaction.

Electroweak unification

The most outstanding fundamental advance of physics in the nineteenth century was undoubtedly the unification of electricity and magnetism as expressed in Maxwell's electromagnetic equations. The solutions to these equations proved conclusively that light-waves, and radio-waves and gamma-rays, are electromagnetic in character. Einstein's theory of special relativity combined with quantum mechanics led to relativistic quantum mechanics, which is the basis for all the significant advances in particle physics theory. In particular, quantum field theory was made possible, and in quantum electrodynamics (QED) became the most complete and accurate theory so far encountered for electromagnetism at all distances and all energies. This achievement was in spite of the need to develop techniques, known as renormalization, to subtract out infinities of mass and charge generated by the particular model (electrons as points) and the particular physical processes considered (vacuum polarization by virtual e^+e^- pairs).

At the heart of QED is the use of perturbation theory whereby the effect under study is divided into two parts, an easy part that can be treated exactly and constitutes the major part of the result, and a smaller part which includes the difficulties. This 'perturbation' can be approximated by the use of series expansions in powers of a parameter related to the strength or coupling constant of the interaction. For QED this parameter is the fine structure constant, $\alpha = \mu_0 e^2 c/2h = 1/137$, and therefore becomes very small for powers higher than that in α^3. But for the strong nuclear interaction the corresponding parameter is near unity, and the series does not converge. It might at first be considered that the weak interaction would be even easier than the electromagnetic since its coupling constant is much smaller than $1/137$, but it imports other problems including the need for field quanta with electric charge and mass, and also parity non-conservation.

Fermi's theory of beta-decay, for example $n \rightarrow p + e^- + \bar{v}_e$, assumed the four particles involved interacted at a point. For over 40 years since 1934 it was in general terms successful, apart from neglecting parity non-conservation. It was applied to the decay of the muon, $\mu^+ \rightarrow e^+ + v_e + \bar{v}_\mu$, and the corresponding coupling constant could be related to that for beta-decay in terms of leptonic parts and hadronic parts of the weak interaction, with the latter separated into strangeness-changing and non-changing processes, as encountered with hyperon decays. Nevertheless, Fermi's theory is fundamentally defective since it leads to problems of divergence. Thus it gives the cross-section for the scattering of neutrinos by electrons as proportional to the neutrino energy, E_v, such that at about $E_v = 6 \times 10^8$ GeV the total probability is greater than unity, i.e. 'unitarity' is violated.

The original suggestion of Yukawa in 1938 was therefore resurrected, replacing the four-fermion interaction at a point by an exchange of a field boson of finite mass, the W boson. An extension of the Fermi interaction involving the W boson still leads to problems of divergence at high energies. At low energies it requires a mass of about 80 GeV/c^2, and by Yukawa-type arguments this implies an extremely short range, $R = \hbar c/Mc^2$ (see p. 126), with R about 10^{-18} m. Furthermore, the higher-order terms from perturbation theory are all infinite and cannot be eliminated by renormalization techniques. The success of renormalization in QED arises from the fact that it is a 'gauge theory' (see p. 162) and this approach became increasingly promising.

As early as 1961, Glashow identified the requisite conserved quantities for the weak interaction and hence the underlying symmetry. The grouping accommodated not only the charged W's, but also two uncharged particles (Glashow 1961). One was identified with the photon and was the first suggestion of an underlying link between the weak and the EM interactions. The W^+ and W^- are required for such reactions as

$$v_\mu + n \rightarrow p + \mu^-,$$

and a neutral particle, now termed the Z^0, is required for reactions such as the elastic v–e reactions mentioned above. The 'flow' of particles into and out of an interaction region, containing the linking 'gauge boson' exchange, is usually referred to as a 'current', by analogy with the flow of electric charge involved in QED. We therefore need to consider both charged-current and neutral-current cross-sections, and at the time when the Z^0 was proposed the latter had not been observed. In 1973, experiments at CERN showed that neutrinos interacting with protons need not always convert to charged leptons, but could retain their charge identities and be elastically scattered by a neutral current process (Hasert *et al.* 1973). This observation was particularly encouraging to theorists who had begun to recognize that gauge theory would not only provide a firm basis for the weak interaction, but would also unify the weak and electromagnetic interactions (Weinberg 1964).

A serious problem was the origin of the masses of the W's and Z^0, which were expected to be zero in any gauge theory, as is the photon in QED gauge theory. A mechanism for endowing mass to the W's and Z^0 was recognized by Higgs (1964) and incorporated by Salam (1968), and independently by Weinberg (1967), into the symmetry scheme of Glashow. The further problem of renormalization of the resulting gauge theory was solved by t'Hooft (1971) and the 'Standard Model' was then ready to make some very precise predictions amenable to test by experiment, such as the masses of the W^+, W^-, and Z^0, their very short lifetimes ($\sim 10^{-24}$ s), and their decay schemes, for W^- for example 8 per cent by $e^- \bar{\nu}_e$, and for Z^0 3 per cent by $e^+ e^-$. The theory also implies that there should be a massive 'Higgs particle'.

The experimental discovery of the W^+, W^-, and Z^0 at CERN in 1983 (see pp. 86 ff), with masses very close to those predicted (W's at 82.3 GeV/c^2, and Z^0 at 92.8 GeV/c^2), has been recognized as confirmation of the theory unifying the electromagnetic and weak interactions.

Tauons and bottom and top quarks

Amidst the excitement of the discovery of the J/ψ and its confirmation of the 'charm' hypothesis, another remarkable particle was found in $e^+ e^-$ colliding beam studies at Stanford (Perl *et al.* 1975). This was the tauon (τ), a heavy lepton of mass 1.780 GeV/c^2, with a lifetime of about 5×10^{-13} s. Its decay into leptons, either e or μ, with neutrinos, establishes its leptonic identity and it is therefore produced in pairs:

$$e^+ + e^- \rightarrow \tau^+ + \tau^-$$
$$\mu^- \bar{\nu}_\mu \nu_\tau$$
$$e^+ \nu_e \bar{\nu}_\tau$$

Other observed decays include up to three pions. The energy spectra of the decays, and the variation of the cross-section for production near the energy threshold, favour a spin of $\frac{1}{2}$. The lifetime indicates that it is associated with its own neutrino ν_τ, and we therefore assign a new lepton number L_τ and recognize that there are three 'generations' of leptons, (e, v_e), (μ, v_μ), (τ, v_τ).

The consequences of the discovery of the tauon are considerable. Since lepton–quark symmetry arguments led successfully to the discovery of the flavour charm, a third pair of leptons implies that there should be a third pair of quark flavours. These are now referred to as 'bottomness' (or 'beauty') and 'topness' (or 'truth') and the new top quarks and bottom quarks are available for a whole new spectrum of composite particles. Omitting spin and colour, the appropriate groups rise in rank from SU(4) to SU(6), and diagrams such as Figs. 8.5–8.7 in two dimensions and Fig. 9.2 in three dimensions, need to be extended to five dimensions. We assign new quantum numbers to parallel S for strangeness and C for charm, and choose B^* for bottomness (the star distinguishing it from baryon number B), and T for topness (encouraging us to use I and I_3 for isospin!). The equation for the charge of a particle then becomes

$$Q = I_3 + \tfrac{1}{2}(B + S + C + B^* + T).$$

The first evidence of hadrons containing b quarks came from 400 GeV protons incident on nuclei at the Fermi Laboratory (Herb *et al.* 1977), at masses in the 9–10 GeV/c^2 region. A cleaner study, free from the extra 'spectator' particles present in hadron collisions, was made at Cornell using e^+e^- and four so-called upsilon (Υ) states were observed at 9.456, 10.016, 10.347 and 10.570 GeV/c^2. The first three are 'narrow' in energy uncertainty corresponding to lifetimes of about 2×10^{-20} s, and the last with a width of ~ 20 MeV has a lifetime of about 3×10^{-23} s. These are interpreted as $(b\bar{b})$ mesons, 'bottomonium', parallelling the J/ψ ($= c\bar{c}$, charmonium), the heaviest of the four decaying to two mesons with explicit bottomness, since it has just sufficient mass to do so. These b-flavoured mesons with the predicted mass, ~ 5.2 GeV/c^2, have been confirmed (Behrends *et al.* 1983) and decay primarily to charmed mesons. The B^- mass is 5279.3, and the $\bar{B}^0$ mass is 5281.3 MeV/c^2. It is concluded that the effective mass of the b quarks is $\gtrsim 5$ GeV/c^2. The lifetimes are $\sim 10^{-13}$ s for the B^- and 5×10^{-13} s for the $\bar{B}^0$.

Hadrons with top quarks are the next particles to be sought, but there are no reliable predictions of their masses. Searches up to 45 GeV total energy in e^+e^- collisions produced no candidates, and the ratio R (see p. 168) relating the probabilities of hadron production and lepton production in such collisions did not change significantly.

For t, the second of the new pair of quarks, R would be expected to exhibit step increase of about $\times 1.5$ at the threshold energy, similar to the step observed for the charm threshold (about $\times 1.7$). It appears therefore that 'toponium' has a mass greater than 45 GeV/c^2. CERN p$\bar{\text{p}}$ results suggest a top quark effective mass >45 GeV/c^2.

In Table 9.3 the properties of the six quarks are summarized. Each occurs in three colours, making a total of eighteen distinct quarks. Lepton pairs in three generations give a total of six leptons. (Recent p$\bar{\text{p}}$, e$\bar{\text{e}}$ studies imply <5 lepton generations.) The field bosons total twelve—the photon, W^+, W^-, Z^0, and eight different gluons. Then there are the as yet unidentified Higgs particle and the graviton. That suggests a total of 38 'elementary', structureless, point-like particles and there is also the magnetic monopole predicted by Dirac (1931). It seems a larger number than we might hope for, and raises several questions; 'Is there a smaller number of constituent particles of which (some of) these 38 are composed?'. 'And if there is, are they, like the quarks, only indirectly observable?', or perhaps, 'Through insufficient energy in accelerator beams on Earth, are they observable (indirectly) only in galactic or cosmic events?'

TABLE 9.3

Properties of the six quarks

	B	I_3	Q	S	C	B^*	T	Effective mass in GeV/c^2
u	$\frac{1}{3}$	$\frac{1}{2}$	$\frac{2}{3}$	0	0	0	0	0.336
d	$\frac{1}{3}$	$-\frac{1}{2}$	$-\frac{1}{3}$	0	0	0	0	0.336
s	$\frac{1}{3}$	0	$-\frac{1}{3}$	-1	0	0	0	0.5–0.7
c	$\frac{1}{3}$	$\frac{1}{2}$	$\frac{2}{3}$	0	$+1$	0	0	1.5–2.0
b	$\frac{1}{3}$	-1	$-\frac{1}{3}$	0	0	$+1$	0	$\gtrsim 5$
t	$\frac{1}{3}$	0	$\frac{2}{3}$	0	0	0	$+1$	>45

Particle physics and cosmology

The comprehensive progress of particle and nuclear physics has provided new insights in cosmology, in particular in our understanding of the origin and evolution of the matter of the universe. In turn, the observations and theories of astronomy and cosmology have indicated possible constraints on particle physics theories, and have enhanced the importance of certain experimental programmes in particle and nuclear physics.

It is now accepted that the universe began with a Big Bang $\sim 1.5 \times 10^{10}$ years ago. Evidence of the expanding universe provided by the Doppler shifts towards the red in star spectra is summarized in Hubble's Law:

$$\text{receding velocity} = \text{Hubble's Constant} \times \text{distance},$$

with Hubble's Constant (H) about $100 \text{ km s}^{-1} \text{ Mpc}^{-1}$, and 1 Mpc (megaparsec) about 3 million light years or 3×10^{22} m. If we ignore the decrease in velocity produced by the action of self-gravity, Hubble's Law implies that 10^{10} years ago all the galaxies were very close to each other, and in the limit the density was infinite and the radius zero, the initial singularity.

Penzias and Wilson (1965) discovered a uniform background of microwaves corresponding to black-body radiation with a temperature of about 3 K, peaking at about 1 mm wavelength. Since an expanding universe is associated with the cooling of such radiation, we can work backwards in time and calculate that the temperature one second after the Big Bang was about 10^{10} K, and at zero time the temperature was formally infinite. It was at times of the order of a second that nuclear fusion of protons and neutrons led to the formation of the dominant light element nuclei, namely ^{2}H, ^{3}He, ^{4}He, and ^{7}Li. Knowledge of the cross-sections for these processes at energies corresponding to the temperatures of the model leads to predictions of relative abundancies that agree well with those observed.

At such high temperatures, the photons are energetic enough to create particle pairs, such as e^+e^- and $p\bar{p}$, which rapidly interact and come to thermal equilibrium with the radiation field. The weak interaction produces neutrino pairs, e.g. $e^+ + e^- \rightarrow \nu_e \bar{\nu}_e$, and in the early stages of the process these too would be thermalized. But at an age for the universe of about one second the neutrino interactions become so small that their thermalization ceases, and they are frozen out with a density that allows the calculation of the neutrino density today,

$$n_\nu \sim 10^8 \text{ m}^{-3},$$

about one quarter of the present 3 K photon density. From observations of stars and estimates of interstellar gas, the proton density today is deduced to be

$$n_p \sim 0.3 \text{ m}^{-3},$$

which agrees with theories of primordial nucleosynthesis of the light elements. Hence, allowing for the three types of neutrinos, we have

$$n_\nu \sim 10^9 \, n_p,$$

and the significance of a possible non-zero mass for neutrinos soon becomes evident.

The future of the universe depends on its total mass. Gravitational attraction to the centre of the spherical distribution of matter slows down the expansion, and if this factor is large enough, i.e. if the mass of the universe is large enough, the expansion will be halted at a certain radius. Thereafter it will collapse back to a point, the so-called Big Crunch. Calculations, using the observations and arguments already mentioned, indicate that if the mass of a neutrino is about 15–30 eV/c^2 then the mass of the universe is near the critical value between that for perpetual expansion with a freezing to death and that for an eventual 'heat death' in the Big Crunch.

Measurements of the mass of the electron antineutrino are clearly of major importance. Workers in Moscow have reported values between 20 and 45 eV/c^2 (Fritschi et al. 1986). This has not been confirmed by recent measurements at Zurich which result in values less than 18 eV/c^2 that are considered not to be inconsistent with zero rest mass. The experiments use magnetic spectrometers to study the energy spectrum of β's from tritium decay, and require accurate determinations of the maximum β-energy—a very difficult task since the sources need to be thin to minimize errors in corrections for energy loss.

On 23 February 1987 the brightest supernova for nearly four hundred years became easily visible to the unaided eye, in the Large Magellanic Cloud of the southern hemisphere. The timescale for the initiating collapse of such a star is less than one second, and all three types of neutrino, electron-, muon-, and tauon-neutrinos, are produced in about equal numbers ($\sim 10^{57}$) in temperatures as high as 10^{10} K. About 10 per cent of the electron neutrinos are produced, together with neutrons, by proton capture of electrons. The flux of neutrinos at the Earth on that day was about 10^{12} per square centimetre. About 30 of these were detected in three observatories— in Japan (Hirata et al 1987), in the USA (Bionta et al 1987) and in the

USSR (Pomansky 1987)—in apparatus that has been recording neutrinos from the Sun at the rate of one or two per day. Since the neutrinos travelled $1.6 \; 10^5$ years their lifetime is at least of the order $10^5 \; m_v c^2/(m_v c^2 + E)$ years, ~ 40 days. (So the 'shortage' of solar v's is not due to decay.) Their time of arrival is given by

$$t = d/v = d/c + m_v^2 c^3/2E^2 + \ldots$$

and observed arrival times were spread over about 10 secs. Plots of t against $1/E^2$, for observed energies up to $E \simeq 40$ MeV, suggest an upper limit to the mass of the neutrinos detected of about $10 \; \text{eV}/c^2$, i.e. not massive enough to close the Universe. These analyses depend significantly on theoretical estimates of the time distributions of the formation and diffusion of the neutrinos in the supernova (Burrows and Lattimer 1987).

It is interesting to note that the neutrino mass must be less than about $100 \; \text{eV}/c^2$, otherwise the mass of the universe would be so large that it would have a 'life', before expansion turned to contraction, shorter than the known age of objects such as the Earth and Sun. Such a mass limit estimate depends, of course, on the validity of the cosmological models and the accuracy of the observations from astronomy used in their computations.

Progress in particle physics has been closely linked to increases in the energy available in centre of mass in particle collisions, and on developments in detectors to acquire the requisite data for sophisticated analysis. Such increases in energy are equivalent to returning to earlier and earlier times in the development of the Big Bang. Thus we can consider times, or ages, corresponding to energy thresholds for the production of not only $p\bar{p}$ pairs but the more exotic particles such as the J/ψ meson ('charmonium') and the τ-lepton, and with some confidence include not only 'bottom' particles but the so-far undiscovered 'top' particles. It has hitherto been found particularly valuable in the laboratory to use 'clean' $e^+ e^-$ collisions, and to a lesser extent those from $p\bar{p}$ and ep, but as noted earlier (see p. 103) the collision of heavy nuclei at 100 GeV per nucleon energies will create conditions similar to those when the universe was only microseconds old.

The unification of the forces of electromagnetism and the weak interaction becomes manifest at high enough energies, as evidenced by the detection of W^+, W^-, and Z^0, with rest masses of about 80 and $90 \; \text{GeV}/c^2$. Such successes stimulate attempts to unify the electroweak with the strong interaction and invites the use of SU(5) group theory linking the SU(2) × U(1) of the Standard Model of the electroweak with the SU(3) of the strong. Such grand unified theories are referred to as GUTs and a number of attempts have been made, with several common features. Thus it appears that the three so-called 'running' coupling constants become equal at an energy corresponding to a unification mass of about 10^{15} GeV/c^2—far beyond the reach of any current accelerator techniques. The characteristics of such high energies can be predicted, however, and in particular the lepton–quark links imply that protons ought to decay, for example to $\pi^0 e^+$. The predicted lifetimes for the several modes of decay depend on the particular GUT but are of the order 10^{31} years.

Several proton-decay experiments are in progress but there is, as yet, no satisfactory evidence of the decay. A volume of 1000 tonnes of iron is needed to provide enough protons to produce on average one $\pi^0 e^+$ decay per year if the lifetime is about 10^{31} years. The experiments are therefore very difficult, especially as the decay products need to be identified against a significant background of neutrino and muon events. The best results to date indicate a lifetime in excess of 10^{32} years, and this is inconsistent with the present predictions from SU(5) theory.

There is an excess of matter over anti-matter in the observed universe and it is an attractive aspect of most GUTs that they can explain this, but only at the expense of also predicting proton decay. In the same context, however, they can explain the ratio of baryons to photons, which is about 10^{-9}. Another outcome of GUTs is the identity of the magnitudes of the proton and electron charges.

The fourth familiar interaction is gravity and to synthesize gravitational and quantum effects it is expected that energies about 10^{19} GeV need to be considered. In terms of the uncertainty principle, this corresponds to times of the order of 10^{-44} s, and distances of the order of 10^{-35} m. There are many attempts to produce a TOE ('theory of everything'), including so-called 'supersymmetric' theories. These are based on theories that cease to treat bosons and fermions as distinct, and instead treat them as equal within a defined symmetry. The graviton of spin 2 is then partnered by a 'gravitino' of spin $\frac{3}{2}$, for example. At lower energies, 10^2–10^{15} GeV, such supersymmetry theories provide leptons and quarks with boson partners ('selectrons', 'squarks', etc.) and field bosons have fermion partners ('photinos', 'gluinos'). Whereas earlier studies suggested an absence of new particles with rest masses in this energy-range, the so-called 'desert region', these recent ideas may provide many new particles to keep the experimentalists busy.

The infinities in relativistic quantum field theories are known to arise from the point-like characteristics of the particles. An entirely new approach replaces the point by the simplest kind of extended object, namely a string. It is possible to treat such systems in 10 space–time dimensions and these 'superstring' theories are considered by some to include the sort of radical rethinking needed to provide the next generations of advances in understanding and hence testable predictions. 'Supermembranes' in 11 dimensions may follow!

One thing is certain, this Earth is too small to allow us to generate the energies and conditions needed to test the particle theories that relate to the very earliest moments of creation. Particle physicists, astronomers, and cosmologists are here on common ground, at the very frontiers of fundamental physical science.

Problems

9.1. Which of the following reactions and decays are 'forbidden' by one or more conservation laws? State the laws violated.

(a) $n \rightarrow p + 2e^- + e^+ + \nu_e$.

(b) $\mu^+ \rightarrow e^+ + \gamma + \bar{\nu}_e$.

(c) $\pi^0 \rightarrow \mu^- + e^+$.

(d) $\pi^- \rightarrow \mu^+ + 2e^- + \nu_\mu$.

(e) $p + p \rightarrow \Sigma^+ + K^+$

(f) $n + d \rightarrow \Sigma^0 + \Sigma^0 + p$.

(g) $K^- \rightarrow \pi^- + \mu^+ + \mu^-$.

(h) $K^+ \rightarrow e^+ + \pi^0 + \nu_e$.

(i) $\pi^+ + n \rightarrow \Lambda^0 + K^+$.

(j) $K^- + p \rightarrow \Xi^0 + K^0$.

9.2. The ratio R (see p. 168) is given by $\Sigma_i Q_i^2$ with Q_i the electric charges of the i quarks. If there are 3 colours, R is three times larger. Calculate R for (a) no colour and (b) 3 colours, for 3, 4, 5, and 6 quarks and hence sketch R as a function of energy, assuming no other processes influence R.

9.3. Identify the following $J^\pi = \frac{1}{2}^+$ baryons from their quark constituents; uud, uus, uss, dds, dss and udd.

9.4. Identify the following $J^\pi = 0^-$ (i.e. pseudoscalar) mesons from their quark constituents; $u\bar{d}$, $u\bar{s}$, $d\bar{s}$, $\bar{d}s$, $\bar{u}d$, $\bar{u}s$.

9.5. Calculate the numbers of (a) baryon states and (b) meson states with distinctive quark compositions for 3, 4, 5, and 6 quarks. (NB. these numbers are much smaller than the numbers of distinct particles since the quark spins can combine in a variety of ways, there can also be orbital angular momentum, and some particles are mixed states.)

Appendix A: Rutherford scattering theory

Consider a nucleus of mass M, electric charge Ze, acting as a stationary scattering centre for a parallel beam of particles each of mass m, charge ze, and momentum p. For simplicity we shall at first assume $M \gg m$ so that nuclear recoil can be neglected and all the angles can be taken as angles in the laboratory system.

Let the impact parameter d(Fig. A.1) lead to a scattering angle θ, and consider the incident particle at point A. The diagram is clearly symmetric about the line NS since the motion of the particle can be reversed without affecting the result. Let φ be the angle between NS and NA. The force F on the particle at point A is equal in magnitude to $(1/4\pi\varepsilon_0)\, zZe^2/r^2$, where $r = $ NA. The change of linear momentum during the complete scattering process amounts to $\delta p = 2p\sin(\theta/2)$ parallel to NS, as can be readily deduced from the vector diagram (Fig. A.1), and is produced by the component of F parallel to

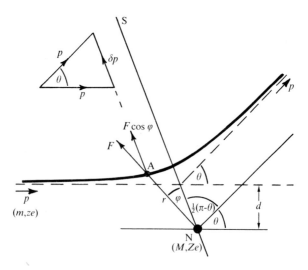

Fig. A.1. Parameters in the Coulomb scattering of a charged particle A by a nucleus N. The impact parameter is the distance d.

NS, i.e. $F \cos \varphi$. Hence

$$\int_0^\infty \left(\frac{1}{4\pi\varepsilon_0}\right) \frac{zZe^2}{r^2} \cos \varphi \, dt = 2p \sin(\theta/2).$$

Since angular momentum is conserved,

$$pd = m\left(r \frac{d\varphi}{dt}\right)r,$$

and this allows the elimination of r, yielding

$$\frac{zZe^2}{4\pi\varepsilon_0} \int_0^\infty \frac{m}{pd} \cos \varphi \frac{d\varphi}{dt} \, dt = 2p \sin(\theta/2).$$

The limits of φ are $-\frac{1}{2}(\pi - \theta)$ and $+\frac{1}{2}(\pi - \theta)$,
hence

$$\frac{zZe^2}{4\pi\varepsilon_0} \cdot \frac{m}{2p^2d} \int_{-\frac{1}{2}(\pi-\theta)}^{\frac{1}{2}(\pi-\theta)} \cos \varphi \, d\varphi = \sin(\theta/2).$$

It is convenient to introduce b_0, the distance of closest approach for $\theta = 180°$
(Fig. 3.3), given by

$$b_0 = \frac{zZe^2}{4\pi\varepsilon_0} \cdot \frac{2m}{p^2}$$

Then, by integration,

$$\cot(\theta/2) = 2d/b_0.$$

As explained on p. 36, the differential scattering cross-section is the
probability of scattering by one nucleus per unit area into unit solid angle in
the direction θ. Consider therefore a range of the impact parameter $+\delta d$,
which will produce a range of $-\delta\theta$ (Fig. A.2); notice that an increase of d leads

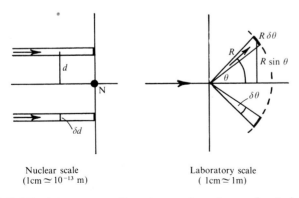

Nuclear scale
$(1\text{cm} \simeq 10^{-13} \text{ m})$

Laboratory scale
$(1\text{cm} \simeq 1\text{m})$

Fig. A.2. Relation between range of impact parameter and range of scattering angle.

to a decrease of θ. For one nucleus per unit area, the probability of scatter into $-\delta\theta$ is clearly the ratio of the area $2\pi d \cdot \delta d$ to unity. The solid angle corresponding to $-\delta\theta$ is $2\pi R \sin\theta(-R\delta\theta)/R^2$, i.e. $-2\pi \sin\theta\delta\theta$. Hence

$$\frac{d\sigma(\theta)}{d\Omega} = \frac{-2\pi d \cdot \delta d}{2\pi \sin\theta\delta\theta} = \frac{-d}{\sin\theta} \cdot \frac{\delta d}{\delta\theta}.$$

From the differentiation with respect to θ of $\cot(\theta/2) = 2d/b_0$,

$$-\frac{1}{2\sin^2(\theta/2)} = \frac{2}{b_0} \frac{\delta d}{\delta\theta},$$

and hence, using $\sin\theta = 2\sin(\theta/2)\cos(\theta/2)$,

$$\frac{d\sigma(\theta)}{d\Omega} = \frac{-(b_0/2)\cot(\theta/2)}{2\sin(\theta/2)\cos(\theta/2)} \left\{ \frac{-b_0/2}{2\sin^2(\theta/2)} \right\}$$

or

$$\frac{d\sigma(\theta)}{d\Omega} = \frac{b_0^2}{16\sin^4(\theta/2)}.$$

Allowance for the finite mass of the nucleus involves two corrections, which are small for low-energy (~ 5 MeV) alpha-scattering from heavy nuclei. The first is the use of the reduced mass in the expression for b_0,

$$b_0 = \frac{zZe^2}{4\pi\varepsilon_0} \cdot \frac{2m}{p^2} \left(\frac{M+m}{M} \right).$$

The second arises from the translation of angles (θ_{CM}) in the centre-of-mass system to angles (θ) in the laboratory system,

$$\left\{ \frac{d\sigma(\theta)}{d\Omega} \right\}_{CM} = \frac{b_0^2}{16\sin^4(\theta_{CM}/2)},$$

with

$$\tan\theta = \frac{\sin\theta_{CM}}{m/M + \cos\theta_{CM}}$$

Notice that in calculating the number scattered,

$$N(\theta_{CM}) = \left\{ \frac{d\sigma(\theta)}{d\Omega} \right\}_{CM} N_0 \, nx \, \delta\Omega_{CM},$$

and since $\delta\Omega = 2\pi \sin\theta\delta\theta$,

$$\delta\Omega_{CM} = \left[\frac{\{1 + (2m/M)\cos\theta_{CM} + m^2/M^2\}^{\frac{3}{2}}}{1 + (m/M)\cos\theta_{CM}} \right] \delta\Omega,$$

with $\delta\Omega = D/R^2$ for a detector of area D at a distance R from the scattering centre.

The angular dependence of the Rutherford scattering formula indicates that $d\sigma(\theta)/d\Omega$ becomes infinitely large as θ tends to zero. However, for θ small the distance of closest approach is so large that the atomic electrons partially shield the nuclear electric charge. The formula therefore needs to be adjusted with the result that the total elastic cross-section σ_{el} becomes finite.

Appendix B: Transmission through a potential barrier

Consider a beam of particles, each of mass M and kinetic energy T incident on a square potential barrier (Fig. B.1(a)) with $V(x)=0$ for $-b>x>0$, and $V(x)=V$ for $-b<x<0$. For α-decay, $T=Q_\alpha$ (p. 48).

In the region $x<-b$ the incident wave can be assigned unit amplitude and is represented by $\exp(i\alpha x)$, and the reflected wave is correspondingly $A\exp(-i\alpha x)$. The total wave in this region is therefore

$$\psi(x)=\exp(i\alpha x)+A\exp(-i\alpha x),$$

and this satisfies the wave equation for $\alpha^2=2MT/\hbar^2$, with the amplitude A as yet unspecified and in general complex.

Within the barrier the form of the wavefunction cannot be that of a free particle, and the 'waves' in the two directions are $B\exp(-\beta x)$ forwards and $C\exp(\beta x)$ in reverse, i.e reflected at $x=0$. Then $\beta^2=2M(V-T)/\hbar^2$, and B and C are in general complex. Hence for $-b<x<0$

$$\psi_2=B\exp(-\beta x)+C\exp(\beta x).$$

The transmitted wave is simply

$$\psi_3=D\exp(i\alpha x),$$

where D is also in general complex and the probability of transmission is $|D|^2$.

The boundary conditions require continuity of both ψ and $d\psi/dx$ at $x=-b$ and $x=0$, and it follows that we have four equations for the four complex unknowns.

At $x=0$,
$$B+C=D,$$
$$-\beta B+\beta C=i\alpha D.$$
At $x=-b$,
$$\exp(-i\alpha b)+A\exp(i\alpha b)=B\exp(\beta b)+C\exp(-\beta b)$$
$$i\alpha\exp(-i\alpha b)-A i\alpha\exp(i\alpha b)=-B\beta\exp(\beta b)+C\beta\exp(-\beta b).$$

For the present purposes only D is needed. Successive elimination of A, B,

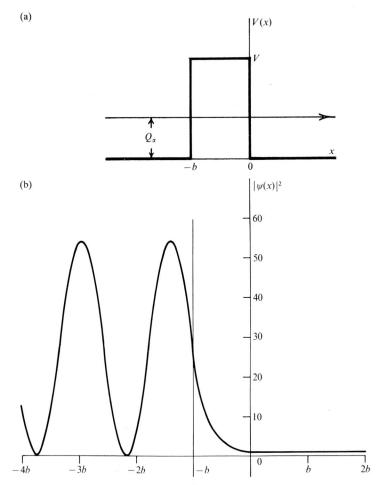

Fig. B.1. (a) Square barrier used in simple theory of alpha-emission. (b) corresponding probability per unit length, $|\psi(x)|^2$, in the three regions, calculated for $Q_\alpha = V/2$, unit transmitted intensity, and $b = 2/\alpha$, i.e. a barrier width equal to twice the reduced wavelength ($\lambda/2\pi = 1/\alpha$) of the incident wave. For $Q_\alpha = V/2$ and $b = 2/\alpha$, unit transmitted intensity requires an incident amplitude of $1/|D| = 3.762$.

and C yields, after a little tedious algebra,

$$D = 2i\alpha \exp(-i\alpha b)\left[\frac{\alpha^2 - \beta^2}{2\beta}\left\{\exp(\beta b) - \exp(-\beta b)\right\}\right.$$
$$\left. + i\alpha\left\{\exp(\beta b) + \exp(-\beta b)\right\}\right]^{-1}.$$

The transmitted intensity is then

$$|D|^2 = \left(\frac{4\alpha\beta}{\alpha^2 + \beta^2}\right)^2 \{\exp(2\beta b) + \exp(-2\beta b) + 2\}^{-1}$$

or

$$|D|^2 = \left(\frac{2\alpha\beta}{\alpha^2 + \beta^2}\right)^2 \{1 + \sinh^2(\beta b)\}^{-1}.$$

Inserting the values of α^2 and β^2 leads to

$$|D|^2 = 4(T/V)(1 - T/V)\{1 + \sinh^2(\beta b)\}^{-1},$$

with βb for alpha-particles given by

$$\beta b = 0.4365 b (V - T)^{\frac{1}{2}}$$

for b in fm and V and T in MeV.

In Fig. B.1(b) we show the probability $|\psi(x)|^2$ of finding the alpha-particle in unit length for the three regions, before, within and after the barrier. The interpretation of Fig. B.1(b) is not difficult. The oncoming wave, $\exp(i\alpha x)$ with $\alpha^2 = 2\mu_\alpha Q_\alpha/\hbar^2$, is partially reflected and partially transmitted at the front of the barrier ($x = -b$). The 'wave' that enters the barrier is partially reflected at the far side of the barrier ($x = 0$) and hence multiple partial reflections take place at $x = -b$ and $x = 0$. The sum of the reflected waves for $x < -b$ is $A \exp(-i\alpha x)$ and combines with the incident wave to produce a standing wave and therefore a sinusoidally varying intensity superimposed on a finite flux. The 'waves' within the barrier have the forms $\exp(-\beta x)$ and $\exp(\beta x)$ with $\beta^2 = 2\mu_\alpha(V - Q_\alpha)/\hbar^2$. Their sum yields a probability of penetrating the barrier which decreases steadily (not exponentially) with the depth of penetration until the width of the barrier is traversed. The sum of the transmitted waves ($D \exp(i\alpha x)$) corresponds to a steady beam, i.e. a constant probability as a function of x. The flux throughout is clearly independent of x, otherwise there would be a local pile-up or absorption of particles.

Using the simplifying assumptions of p. 48, we can derive for the square-well potential a relation equivalent to that of Geiger and Nuttall (p. 49). But first it is instructive to consider the more realistic potentials that have been proposed, in particular the Coulomb potential cut-off at the nuclear radius by a square well (Fig. 3.9), and also the potential deduced by Igo (1958) for $r > R$ from optical model fits to the differential cross-sections, $d\sigma(\theta)/d\Omega$, for alpha-particle scattering (Fig. 3.9). To obtain an estimate of βb for the square barrier, we note that suitable values of V and b equivalent to the Coulomb plus square-well and also to the Igo potential, are shown by Fig. 3.9 to be about 15 MeV and 30 fm respectively for $Q_\alpha = 7.30$ MeV, and for these values $\beta b = 36$. For such large values of βb, it is a good approximation to replace $(1 + \sinh^2 \beta b)^{-1}$ by $4\exp(-2\beta b)$. We now make the very crude and crucial assumptions, also suggested by Fig. 3.9, that $V \simeq 2Q_\alpha$ with $b \simeq b_0$, where b_0 is the distance of closest approach (Fig. 3.3). Since $b_0 = zZe^2/4\pi\varepsilon_o Q_\alpha$ with $z = 2$

for the alpha particle and Z the atomic number of the residual nucleus after the alpha emission, and $\beta = (2\mu_\alpha/\hbar^2)^{\frac{1}{2}}(V - Q_\alpha)^{\frac{1}{2}}$, it follows that $2\beta b = KQ_\alpha^{-\frac{1}{2}}$ where $K = (8\mu_\alpha)^{\frac{1}{2}} zZe^2/4\pi\varepsilon_0\hbar$. The transmission with these assumptions then reduces to $4\exp(-KQ_\alpha^{-\frac{1}{2}})$.

The radioactive constant λ is the probability of transmission multiplied by the frequency v of impact on the barrier of the N_α alpha-particles in the nucleus. We shall assume that v is independent of Q_α and mass number A. It then follows that λ is approximately proportional to $\exp(-KQ_\alpha^{-\frac{1}{2}})$ or

$$\ln\lambda = A_2 Z Q_\alpha^{-\frac{1}{2}} + B_2,$$

where $A_2 = -K/Z$ and B_2 are approximately constant. This relation is closely equivalent to that of Geiger and Nuttall

$$\ln\lambda = A_1 \ln Q_\alpha + B_1$$

for the range of Q_α encountered, as may be seen in Fig. 3.10. Comparison with experimental values for the three radioactive series yields a fairly good straight line for $\ln\lambda$ plotted against $ZQ_\alpha^{-\frac{1}{2}}$ with $A_2 = -3.93$ and $B_2 = 130$ for λ in s^{-1} and Q_α in MeV.

A value of the frequency of impact v must be assumed in order to calculate values of the lifetimes τ and the constant B_2. It is found that $v = 10^{21}\,s^{-1}$ produces fair agreement with observed lifetimes, and such a frequency corresponds to a pre-formed alpha-particle oscillating across the nuclear diameter with a velocity of about $c/30$. The square-well potential then gives $A_2 = -K/Z = -2.50$ and $B_2 = \ln(4N_\alpha v) = 50$ (for $N_\alpha = 1$). Hence the dependence on Z and $Q^{-\frac{1}{2}}$ appears to be well founded but the many and severe approximations produce significantly different values of A_2 and B_2.

The forms of the wavefunctions in the three regions $x < -b$, $-b < x < 0$, and $x > 0$, are shown in Fig. B.2, which is drawn to emphasize two points (Burge 1970).

1. that it is not possible to represent the amplitude of the wavefunction in all three regions as a purely real quantity;

2. that the forms of the wavefunction within the barrier depend on the phase of the incident wave which has a time-dependence $\exp(-iTt/\hbar)$.

In a wave-mechanical representation of a stationary state the real part of the wavefunction is always of the same form as the imaginary part but, in general, of different amplitude, and it is possible to select a point in time when the imaginary part has zero amplitude, i.e. the wavefunction is purely real throughout the system. Whenever there is a flux of particles, as through a potential barrier, it is not possible to find a point in time at which the wavefunction is purely real (or imaginary) throughout the system, and therefore it is necessary, for a proper interpretation of the wavefunction, to consider both the real and imaginary parts.

The usual diagrams of wavefunctions in potential barriers appear similar to that of the real part of Fig. B.2(a), probably in order to illustrate that the

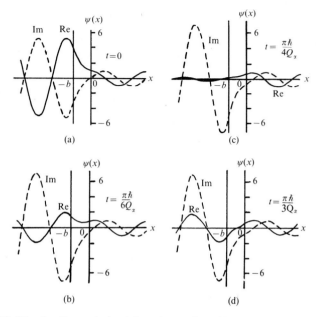

Fig. B.2. Wavefunctions calculated for a beam of particles incident on a potential barrier, at four different times, showing the effect of incident phase on the form of the real (Re) and imaginary (Im) parts of the wavefunction inside the barrier. (The width of the barrier is set at twice the reduced wavelength ($\lambda/2\pi$) of the incident beam; the height of the barrier twice the kinetic energy of the beam; the transmitted intensity is unity; in the expressions for t the kinetic energy is taken as Q_α.) (Burge 1970.)

intensity (i.e. $|\psi|^2$) decreases monotonically within the barrier (Fig. 3.10(b)). But it is clear that $|\psi|^2$ is the same for all the incident phases shown in Fig. B.2, and the amplitude of the real part (or the imaginary part) can have any of the forms shown, for all of which $d^2\psi/dx^2$ is positive within the barrier.

Appendix C: The square-well potential for the deuteron

The solution to the wave equation

$$\frac{d^2 U(r)}{dr^2} + \frac{2\mu}{\hbar^2}[E - V(r)]U(r) = 0$$

is given by $U(r < b) = A \sin Kr$

and $U(r > b) = B \exp(-\alpha r)$

with $K^2 = (2\mu/\hbar^2)(V_0 - B_d)$ and $\alpha^2 = 2\mu B_d/\hbar^2$.

From the boundary conditions

$$K \cot Kb = -\alpha,$$

which can be rearranged to give

$$b = (2\mu/\hbar^2)^{-\frac{1}{2}}(V_0 - B_d)^{-\frac{1}{2}}[\pi + \tan^{-1}\{-[(V_0 - B_d)/B_d]^{\frac{1}{2}}\}]$$

and the form of this relationship between V_0 and b is as shown in Fig. 7.4.

We now use R_{rms} to produce a second relation between b and V_0. The theoretical value of R_{rms} is evident if we consider a triangle CPQ with C the centre-of-mass, P the proton centre and Q the charge element distant $\langle r_p^2 \rangle^{\frac{1}{2}}$ from P. The mean of CQ^2 over all angles CPQ is the average of $CP^2 + PQ^2 + 2CP \, PQ \cos(\pi - CPQ)$, i.e. $CP^2 + PQ^2$, and hence

$$R_{rms}^2 = \langle (r/2)^2 \rangle + \langle r_p^2 \rangle.$$

The mean value of $(r/2)^2$ is given by

$$\langle (r/2)^2 \rangle = \int_0^b (r/2)^2 A^2 \sin^2(Kr)dr$$
$$+ \int_b^\infty (r/2)^2 B^2 \exp(-2\alpha r)dr,$$

where A and B are obtained from the boundary condition

$$A \sin Kb = B \exp(-\alpha b)$$

together with the normalization relation from total probability,

$$1 = \int_0^b A^2 \sin^2(Kr)dr + \int_b^\infty B^2 \exp(-2\alpha r)dr.$$

Integration by parts and some algebra yield

$$R_{rms}^2 = \frac{1}{8\alpha^2} + \frac{2\alpha^2 b^3 + 6\alpha b^2 + 3b}{24\alpha(1 + \alpha b)} - \frac{1}{8k^2} + \langle r_p^2 \rangle$$

from which it follows that

$$V_0 = B_d + (\hbar^2/2\mu)\{[1/\alpha^2 + (2\alpha^2 b^3 + 6\alpha b^2 + 3b)/(3\alpha + 3\alpha^2 b)]$$
$$- 8R_{rms}^2 + 8\langle r_p^2 \rangle\}^{-1}.$$

A plot of this relation between b and V_0 is shown in Fig. 7.4 for the measured value of R_{rms} and for the two values at the limits of the assigned errors (Burge 1986). From the intersections it follows that the values of b and V_0 that satisfy both B_d and R_{rms} are

$$V_0 = 34.6^{+10.3}_{-6.5} \text{ MeV} \quad \text{and} \quad b = 2.07 \pm 0.30 \text{ fm}.$$

The sensitivity of V_0 and b to the errors assigned to R_{rms} arises from the fact that the probability of an n–p separation greater than b is large, ranging from 0.69 to 0.59 for R_{rms} limits of 2.04 fm and 2.16 fm respectively.

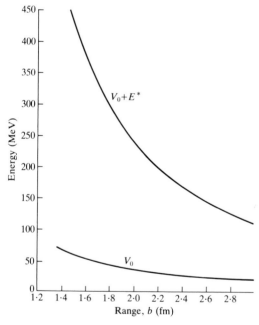

Fig. C.1. Variation of V_0 and $(V_0 + E^*)$ with b for $B_d = 2.225$ MeV, indicating the large positive value of E^*, the energy of the (unbound) first excited state (Burge 1986).

The nearness of $b = 2.07$ fm to $R_{rms} = 2.1$ fm arises from the cancellation of the two factors noted earlier (see p. 122). First, it should be emphasized that R_{rms} is measured from the centre of mass of the deuteron which is midway between the proton and neutron, whereas b is the n–p separation. Secondly, it should be noted that in the solution of the wave equation the proton is being treated as a point charge whereas it has a root-mean-square charge distribution radius $\langle r_p^2 \rangle^{\frac{1}{2}} = 0.8$ fm. The first factor makes R_{rms} smaller than b, and the second makes R_{rms} larger by about the same amount.

In order to investigate the theoretical possibility of an excited state of the deuteron, we note that the boundary condition, $K \cot Kb = -\alpha$ is satisfied not only by the K value for the ground state but also for the value of K from the next half-cycle of $\cot Kb$. This gives a relation between b and V_0 similar to that provided by B_d, but with π replaced by 2π and $-B_d$ replaced by E^*, where E^* is the energy of the possible excited state,

$$b = (2\mu/\hbar^2)^{-\frac{1}{2}}(V_0 + E^*)^{-\frac{1}{2}}[2\pi + \tan^{-1}\{-[(V_0 + E^*)B_d]^{\frac{1}{2}}\}].$$

The value of $(V_0 + E^*)$ can now be plotted as a function of b as shown in Fig. C.1. On the same graph we plot V_0 as a function of b, as deduced from B_d and already seen in Fig. 7.4, and it is clear that E^*, the difference between $(V_0 + E^*)$ and V_0, is positive and ranges from 261 MeV to 144 MeV for the possible range of b values 1.77 fm to 2.37 fm. Since $-B_d$ corresponds to a bound state, E^* positive and so large implies that the first excited state is unbound for this potential, and by a large amount of energy.

Appendix D: An introduction to SU(2) and SU(3), U-spin and V-spin.

The mathematical formalism suitable for ordinary spin and isospin multiplets is termed SU(2) which stands for 'special unitary group for arrays of size 2×2'. The 2×2 refers to matrix notation for the spin (or isospin) operators used in quantum mechanics and need not concern us in this introductory discussion. We shall develop some of the spin and isospin formalism in order to introduce terminology and ideas relevant to the much more powerful classification systems with higher-rank groupings, in particular SU(3).

The wavefunction for a particle with ordinary spin can be expressed as the product of a function of the space coordinates and a function of the spin

$$\psi(x, y, z, s_z) = \phi(x, y, z) \chi(s_z).$$

For particles of spin $\frac{1}{2}$, $\chi(s_z)$ has two values, one for 'spin up' ($\chi(+\frac{1}{2}) = \alpha$, say) and one for 'spin down' ($\chi(-\frac{1}{2}) = \beta$). We can specify an 'operator' σ_+ which has the effect of converting a spin down wavefunction into a spin-up wavefunction and is termed a 'raising operator'. Similarly, a 'lowering operator' σ_- changes spin up to spin down. The terms 'ladder operator' and 'step operator' are also used for σ_+ and σ_-. As might be expected, σ_+ operating on a spin up produces nothing, and σ_- operating on a spin down likewise nothing. We have here a 'shift operator' in one shift direction, namely the z-direction, for which the value of spin can be measured in ordinary 3-dimensional space.

By a complete analogy we can define raising and lowering operators τ_+ and τ_- for the three-dimensional isospin space in which the third component, T_3, of isospin in the 3-direction can be assigned and differences of T_3 can be measured. Then τ_+ acting on a wavefunction for $T_3 = -\frac{1}{2}$ (a neutron, say) will convert it to $T_3 = +\frac{1}{2}$ (a proton), and τ_- acting on a proton wavefunction will convert it to a neutron. Rather obviously, τ_+ acting on a proton, and likewise τ_- acting on a neutron, will produce nothing, i.e. a zero value for the final wavefunction. Note that these 'shift operators' act only between members of the same spin multiplet. Thus, extending the consideration to $T = 1$ multiplets, the shift operator τ_+ can convert Σ^- to Σ^0, and Σ^0 to Σ^+, but it cannot convert a neutron to Σ^0, or a proton to Σ^+, or indeed either n or p to any one of the three Σ's.

In SU(3) we have 3×3 matrices for the operators and produce the eight operators needed for the allowable group operations. Instead of the two-dimensional wavefunction of spin (the α and β elements in a column matrix) and of isospin (say p and n), we now have three dimensions. In early work the three elements were p, n, and Λ, thereby introducing the additive quantum number of strangeness S (see p. 154) as well as the additive quantum number of isospin, T_3. It is, however, more profitable to use numbers proportional to the hypercharge Y (see p. 154) and T_3 to display the SU(3) symmetry, as shown in Figs 8.5, 8.6, and 8.7.

In SU(3) there are found to be six shift operators, corresponding to the two in SU(2), namely τ_+, τ_- for isospin. The six consist of these same two for the 3-direction or third component of isospin (I-spin), plus two for the 3-direction of so-called V-spin, and two for the 3-direction of U-spin. The three directions of 'shift' are clearly seen in Figs 8.5, 8.6, and 8.7 and are labelled T_3, V_3, and U_3. Notice that U_3 produces a shift along a line of constant electric charge, and perpendicular to it is a charge axis. In other words we could replot Figs 8.5, 8.6, and 8.7 on this basis with axes Q and U_3, rather than Y and T_3. On the other hand, the perpendicular to V_3 is not a simple, well-known quantity and so a diagram of V_3 and its perpendicular axis is not helpful.

We note that SU(3) uses the two properties of isospin (T_3) and hypercharge (Y), and in order to produce displays such as Fig. 8.5 we need to select particles with the same spin (J). If we wish to relate these three characteristics (T_3, Y, J) in a single representation we can extend SU(3) to SU(6), but it is sufficient to use one of its subgroups SU(3) × SU(2). SU(3) symmetry as applied here uses 'internal quantum numbers', but SU(2) applied to ordinary spin uses the 'external quantum numbers' of space–time. This use of SU(6) is therefore valid only if the interactions are spin-independent, which is true only in the non-relativistic region. Furthermore, if full SU(6) is assumed only three of the five additive quantum numbers in SU(6) are used for particle classification.

If we wish to include not only isospin (T_3) and hypercharge (Y) but also the new 'flavour' charm (C) (see p. 165), we need SU(4) of which SU(3) is a subgroup. This is evident from Fig. 9.2 where the cut in the Y, T_3 plane (for $C=0$) is the same as Fig. 8.6 apart from the extra $c\bar{c}$, noting that π^0 is a mixture of the states $u\bar{u}$ and $d\bar{d}$, and η^0 is a mixture of $u\bar{u}$, $d\bar{d}$, and $s\bar{s}$. Extending SU(4) to include spin requires SU(8) with SU(4) × SU(2) as a subgroup. To include the colour of the quarks (u, d, s, c) we would need SU(8) × SU(3). Now that the flavours 'bottomness' (B^*) and 'topness' (T) are needed in any thoroughgoing treatment (see p. 172), we extend to SU(6) × SU(2) if spin is included, and to SU(6) × SU(2) × SU(3) if we are also to take care of colour.

Appendix E: Linear momentum conservation and linear-translation invariance

We commonly accept, and regularly confirm to great accuracy, that the location in space of an experiment does not influence its result. In other words there is a 'symmetry' based on the two locations or, alternatively, it is impossible to determine an absolute position in space. An important consequence of this 'invariance' is that linear momentum is conserved, as may readily be shown.

Consider a system of two particles, of masses m_1 and m_2, at the points x_1 and x_2. Let F_1 be the force acting on particle m_1 due to particle m_2. By Newton's second law,

$$\frac{d}{dt}(m_1 \dot{x}_1) = F_1,$$

or, in terms of the interaction potential energy function,

$$F_1 = -\frac{\partial}{\partial x_1} V(x_1, x_2)$$

hence

$$\frac{d}{dt}(m_1 \dot{x}_1) = -\frac{\partial}{\partial x_1} V(x_1, x_2),$$

which is the equation of motion for particle m_1.

Now impose a linear translation of the system by a distance l in the x-direction, such that

$$x_1 \to X_1 = x_1 + l, \ x_2 \to X_2 = x_2 + l.$$

The new equation of motion is

$$\frac{d}{dt}(m_1 \dot{X}_1) = -\frac{\partial}{\partial X_1} V(X_1, X_2),$$

and since

$$\frac{\partial V}{\partial x_1} = \frac{\partial V}{\partial X_1} \cdot \frac{\partial X_1}{\partial x_1}$$

this reduces to

$$\frac{d}{dt}(m_1 \dot{x}_1) = -\frac{\partial}{\partial x_1} V(x_1 + l, x_2 + l).$$

For an infinitesimal translation,

$$V(x_1 + l, x_2 + l) \doteq V(x_1, x_2) + l\left(\frac{\partial V}{\partial x_1} + \frac{\partial V}{\partial x_2}\right),$$

and hence $(\partial V/\partial x_1 + \partial V/\partial x_2)$ must be zero if the two equations of motion, initial and new, are to be identical. (All potential functions of the form $V(x_1, x_2) = V(x_1 - x_2)$ satisfy this condition, as might be expected since $V(x_1, x_2)$ depends only on the relative coordinates.)

It now follows immediately that

$$\frac{\mathrm{d}}{\mathrm{d}t}(m_1\dot{x}_1) + \frac{\mathrm{d}}{\mathrm{d}t}(m_2\dot{x}_2) = -\frac{\partial V}{\partial x_1} \cdot \frac{\partial V}{\partial x_2} = 0,$$

i.e.
$$m_1\dot{x}_1 + m_2\dot{x}_2 = \text{constant},$$

or the linear momentum in the direction x is conserved.

A similar derivation can be made for conservation of angular momentum based on invariance under rotation through a given angle, i.e. it is impossible to measure an absolute direction in space. Likewise conservation of energy arises from invariance under time-translation, i.e. it is impossible to measure an absolute 'point' in time.

Notice the form of this proof. It starts from Newton's second law in order to define linear momentum and uses the 'observed' linear-translation invariance. The conservation of linear momentum, so defined, can also be proved from Newton's third law. The more general principal of linear-translation invariance is preferable for this purpose, although the derivation above, because it is in terms of conservative forces, is also limited in its generality. The conservation law itself is not so limited.

References

Abraham, M. (1903), *Phys. Zeits.* **4**, 57.

Alston, M. H. and 6 others. (1960). *Phys. Rev. Lett.* **5**, 520.

Alvarez, L. (1946). *Phys. Rev.* **70**, 799A, *see also* Alvarez, L. and 9 others. (1955). *Rev. Sci. Instr.* **26**, 111.

Anderson, H. L. (1932). *Science* **76**, 238.

Anderson, H. L. and 3 others. (1952). *Phys. Rev.* **85**, 936.

Aston, F. W. (1920). *Phil. Mag.* **39**, 449.

Aubert, J. J. and 13 others. (1974). *Phys. Rev. Lett.* **33**, 1404.

Augustin, J. E. and 34 others. (1974). *Phys. Rev. Lett.* **33**, 1406.

Bäcklin, E. (1928). Dissertation, Upsala.

Barnes, V. E. and 32 others. (1964). *Phys. Rev. Lett.* **12**, 204.

Becquerel, H. (1896). *Compt. Rend.* **122**, 420.

Becquerel, H. (1899). *Compt. Rend.* **129**, 1205.

Becquerel, H. (1901). *Compt. Rend.* **133**, 977.

Behrends, S. and 78 others. (1983). *Phys. Rev. Lett.* **50**, 881.

Bethe, H. (1939). *Phys. Rev.* **55**, 434.

Bionta, R. M. and 36 others. (1987). *Phys. Rev. Lett.* **58**, 1494.

Bjorken, B. J. and Glashow, S. L. (1964). *Phys. Lett.* **11**, 255.

Blackett, P. M. S. (1925). *Proc. R. Soc.* **A107**, 349.

Bohr, A. and Mottleson, B. R. (1953). *Dan. Mat. Fys. Medd.* **27**, No. 16.

Bohr, N. (1913). *Phil. Mag.* **26**, 1, 476, 857.

Bohr, N. (1936). *Nature* **137**, 344.

Bothe, W. and Becker, H. (1930). *Zeits. f. Phys.* **66**, 289.

Bragg, W. H. (1904). *Phil. Mag.* **8**, 719.

Brown, R. and 5 others. (1949). *Nature* **163**, 82.

Bucherer, A. H. (1908). *Phys. Zeits.* **9**, 755.

Burge, E. J. (1970). *Phys. Educ.* **5**, 298.

Burge, E. J. (1986). *Eur. J. Phys.* **7**, 138.

Burrows, A. and Lattimer, J. M. (1987). *Astrophys. J.* **318**, L63.

Chadwick, J. (1932). *Proc. R. Soc.* **A136**, 692.

Christenson, J. H., Cronin, J. W., Fitch, V. L. and Turtay, R. (1964). *Phys. Rev. Lett.* **13**, 138.

Christofilos, W. (1952). Unpublished. *See* Blewett, J. W. (1956). *Rep. Progr. Phys.* **19**, 37.

Cockcroft, J. D. and Walton, E. T. S. (1930). *Proc. R. Soc.* **A129**, 477.

Cockcroft, J. D. and Walton, E. T. S. (1932). *Proc. R. Soc.* **A137**, 229.

Courant, E. D., Livingston, M.S. and Snyder, H. S. (1952). *Phys. Rev.* **88**, 1190.

Crookes, W. (1900). *Proc. R. Soc.* **A66**, 409.

Crowther, J. A. (1910). *Proc. Camb. Phil. Soc.* **15**, 442.

Curie-Joliot, I. and Joliot, F. (1932). *Compt. Rend.* **194**, 273.

Dalton, J. (1808). *A New System of Chemical Philosophy.* Manchester, p. 213.

de Broglie, L. (1924). *Phil. Mag.* **47**, 446.

Dewar, J. (1908). *Proc. R. Soc.* **A81**, 280.

Dirac, P. A. M. (1928). *Proc. R. Soc.* **A117**, 610 and **A118**, 351.

Dirac, P. A. M. (1931). *Proc. R. Soc.* **A133**, 60.

Einstein, A. (1905a). *Ann. d. Physik* **17**, 132.

Einstein, A. (1905b) *Ann. d. Physik* **17**, 891.

Ellis, C. D. (1922). *Proc. R. Soc.* **A101**, 1.

Fermi, E. (1934). *Zeits. f. Phys.* **88**, 161.

Fermi, E., and Rasetti, F. (1935). *Nuovo Cim.* **12**, 201.

Franck, J. and Hertz, G. (1914). *Verh. Deut. Phys. Ges.* **16**, 512.

Friedrich, W., Knipping, P. and von Laue, M. (1912). Sitzungsber. Bayer. Akad. Wiss. p. 311, *See also Ann. d. Physik* **41**, 971 (1913).

Fritschi, M. and 5 others. (1986). *Phys. Lett.* **173B**, 485.

Gamow, G. (1928). *Zeits. f. Phys.* **51**, 204.

Gamow, G. (1930). *Proc. R. Soc.* **A123**, 386.

Geiger, H. (1908). *Proc. R. Soc.* **A81**, 174.

Geiger, H. and Marsden, E. (1909). *Proc. R. Soc.* **A82**, 495.

Geiger, H. and Marsden, E. (1913). *Phil. Mag.* **25**, 604.

Geiger, H. and Nuttall, J. M. (1911). *Phil. Mag.* **22**, 613.

Gell-Mann, M. (1952). *Phys. Rev.* **92**, 833.

Gell-Mann, M. (1961). Calif. Inst. Technol. Symp., Lab. Rep. No. 20.

Gell-Mann, M. (1964). *Phys. Lett.* **8**, 214.

Glashow, S. L. (1961). *Nucl. Phys.* **22**, 579.

Glashow, S. L., Iliopoulos, J. and Maiani, L. (1970). *Phys. Rev.* **D2**, 1285.

Gurney, R. W. and Condon, E. U. (1928). *Nature* **122**, 439. *See also Phys. Rev.* **33**, 127.

Haga, H. and Wind, C. H. (1899). *Ann. d. Physik* **68**, 884.

Hahn, O. and Strassmann, F. (1939). *Die Naturwiss.* **27**, 11, 89.

Hallwachs, W. (1888). *Ann. d. Physik* **33**, 301.

Han, M. Y. and Nambu, Y. (1965). *Phys. Rev.* **139B**, 1006.

Hasert, F. and 50 others. (1973). *Phys. Lett.* **46B**, 121.

Haxel, O., Jensen, J. H. D., and Suess, H. E. (1949). *Phys. Rev.* **75**, 1766.

Heisenberg, W. (1932). *Zeits. f. Phys.* **77**, 1.

Herb, S. W. and 15 others. (1977). *Phys. Rev. Lett.* **39**, 252.

Hertz, H. R. (1887). *Ann. d. Physik* **31**, 421.

Hess, V. F. (1913), *Phys. Zeits.* **14**, 610.

Higgs, P. (1964). *Phys. Rev. Lett.* **13**, 508.

Hirata, K. and 22 others. (1987). *Phys. Rev. Lett.* **58**, 1490.

HMSO. (1977). ICRP Publication 26.

HMSO. (1984). *UK Review of Radiation.*

Igo, G. (1958). *Phys. Rev. Lett.* **1**, 72.

Jeans, J. H. (1909). *Phil. Mag.* **17**, 229, 773; **18**, 209.

Kaufmann, W. (1901). *Gott. Nach* **2**. *See also* Heaviside, O. (1902). *Electrician* **48**, 804, 945.

Kerst, D. W. (1940). *Phys. Rev.* **58**, 841.

Lattes, C. M. and Gardner, E. (1948). *Phys. Rev.* **74**, 1236.

Lattes, C. M. G., Occhialini, G. P. S., and Powell, C. F. (1947). *Nature* **160**, 453, 486.

Lawrence, E. O. and Livingston, M. S. (1931). *Phys. Rev.* **37**, 1707. *See also Phys. Rev.* **40**, 19 (1932).

Lee, T. D. and Yang C. N. (1956). *Phys. Rev.* **104**, 254.

Lummer, O. and Pringsheim, P. (1897). *Ann. d. Physik* **63**, 395. *See also Verh. Deut. Phys. Ges.* **1**, 23, 215 (1899).

Marsden, E. (1914). *Phil. Mag.* **27**, 824.

Mayer, M. G. (1948). *Phys. Rev.* **74**, 235. *See also Phys. Rev.* **75**, 1969 (1949).

McMillan, E. M. (1945). *Phys. Rev.* **68**, 143.

Meitner, L. and Frisch, O. (1939). *Nature* **143**, 239.

Mendeleev, D. I. (1869). *Zhurnal Russ. Chimi. Obtchestva* **1**, 60.

Michelson, A. A. and Morley, E. (1887). *Silliman Journ.* **34**, 333, 427. *See also Phil. Mag.* **24**, 449 (1887).

Millikan, R. A. (1909). *Phil. Mag.* **19**, 209.

Millikan, R. A. and Bowen, J. S. (1926). *Phys. Rev.* **27**, 353.

Nagaoka, H. (1904). *Phil. Mag.* **7**, 445.

Nakano, T. and Nishijima, K. (1952). *Prog. Theor. Phys. Japan* **10**, 581.

Nambu, Y., Nishijima, K. and Yamaguchi, Y. (1951). *Prog. Theor. Phys. Japan* **6**, 615, 619.

Neddermeyer, S. H. and Anderson, C. D. (1937). *Phys. Rev.* **51**, 884.

Ne'eman, Y. (1961). *Nucl. Phys.* **26**, 222.

Noddack, I. (1934). *Angew. Chem.* **47**, 653.

Oliphant, M. L. (1943). In *Proc. Phys. Soc.* (1947). **59**, 666.

Oppenheimer, R. (1947). *Phys. Rev.* **71**, 462(T).

Pais, A. (1952). *Phys. Rev.* **86**, 663.

Pais, A. (1953). *Prog. Theor. Phys. Japan* **10**, 457.

Penzias, A. A. and Wilson, R. W. (1965). *Astrophys. Journ. (USA)* **142**, 419.

Perkins, D. H. (1947). *Nature* **159**, 126.

Perl, M. L. and 35 others. (1975). *Phys. Rev. Lett.* **35**, 1489.

Planck, M. (1900). *Verh. Deut. Phys. Ges.* **2**, 237.

Planck, M. (1901). *Ann. d. Physik* **4**, 553.

Pomansky, A. (1987). *Proc. XXII Rencontre de Moriond.* (ed. J. Tran Thank Van). Editions Frontieres.

Powell, C. F. (1947). *See* Lattes, Occhialini, and Powell (1947).

Rayleigh, Lord. (1900). *See* Strutt, J. W.

Rebut, P. H. and Keen, B. E. (1987). *Fusion Tech.* **11**, 13.

Reines, F. and Cowan, C. L. (1953). *Phys. Rev.* **92**, 830.

Rochester, G. D. and Butler, C. C. (1947). *Nature* **160**, 855.

Rosenblum, S. (1929). *Compt. Rend.* **188**, 1401.

Rubbia, C., McIntyre, P. and Cline, D. (1976). *Proc. Aachen Neutrino Conf.* *See also Phys. Lett.* **66B**, 429 (1977).

Rutherford, E. (1899). *Phil. Mag.* **47**, 109.

Rutherford, E. (1906). *Phil. Mag.* **12**, 143.

Rutherford, E. (1911). *Phil. Mag.* **21**, 669.

Rutherford, E. (1919). *Phil. Mag.* **37**, 581.

Rutherford, E. and Andrade, E. N. da C. (1914). *Phil. Mag.* **27**, 854 and **28**, 263 (1914).

Rutherford, E., Chadwick, J., and Ellis, C. D. (1930). *Radiations from Radioactive Substances*, Cambridge University Press, Cambridge.

Rutherford, E. and Geiger, H. (1908a). *Proc. R. Soc.* **A81**, 141.

Rutherford, E. and Geiger, H. (1908b). *Proc. R. Soc.* **A81**, 162.

Rutherford, E. and Robinson, H. (1913). *Phil. Mag.* **6**, 26, 717.

Rutherford, E., Robinson, H., and Rawlinson, W. F. (1914). *Phil. Mag.* **28**, 281.

Rutherford, E. and Royds, T. (1909). *Phil. Mag.* **17**, 281.

Rutherford, E. and Soddy, F. (1902). *Phil. Mag.* **4**, 370, 569.

Rutherford, E. and Wood, A. B. (1916). *Phil. Mag.* **31**, 379.

Sakata, S. (1956). *Progr. Theor. Phys. Japan* **16**, 686.

Salam, A. (1968). *Proc. 8th Nobel Symp., Aspensäsgården.* Almqvist and Wiksell, Stockholm, p. 367.

Schwinger, J. (1953). *Phys. Rev.* **91**, 713, 728. *See also* Luders, R. (1954). *Dan. Mat. Fys. Medd.* **28**, No. 5.

Sloane, D. H. and Lawrence, E. O. (1931). *Phys. Rev.* **38**, 2021.

Snell, A. H. and Miller, L. C. (1948). *Phys. Rev.* **74**, 1217.

Soddy, F. (1913). *See* Rutherford, Chadwick, and Ellis (1930), p. 34.

Stoney, G. J. (1894). *Phil. Mag.* **38**, 418.

Strutt, J. W. (Lord Rayleigh). (1900). *Phil. Mag.* **49**, 539.

Szilard, L. and Chalmers, T. A. (1934). *Nature* **134**, 462, 494.

Thomson, J. J. (1894). *Phil. Mag.* **38**, 358.

Thomson, J. J. (1897). *Phil. Mag.* **44**, 293.

Thomson, J. J. (1899). *Phil. Mag.* **48**, 547.

Thomson, J. J. (1904). *Phil. Mag.* **7**, 237.

Thomson, J. J. (1910). *Proc. Camb. Phil. Soc.* **15**, 465.

Thomson, J. J. (1913). *Rays of Positive Electricity and their Applications to Chemical Analysis.* Longmans Green, London, pp. 112 ff.

t'Hooft, G. (1971). *Nucl. Phys.* **B35**, 167.

van de Graaff, R. J. (1931). *Phys. Rev.* **38**, 1919.

Veksler, V. (1945). *Journ. Phys. USSR* **9**, 153.

Villard, P. (1900). *Compt. Rend.* **130**, 1010, 1178.

von Baeyer, O., Hahn, O., and Meitner, L. (1911). *Phys. Zeits.* **12**, 273.

Watters, H. J. (1956). *Phys. Rev.* **103**, 1763.

Weinberg, S. (1964). *Phys. Rev. Lett.* **13**, 168.

Weinberg, S. (1967). *Phys. Rev. Lett.* **19**, 1264.

Wideröe, R. (1928). *Arch. Electrotech.* **21**, 387.

Wigner, E. P. (1937). *Phys. Rev.* **51**, 106.

Wilson, C. T. R. (1900). *Phil. Trans.* **193**, 289.

Wilson, C. T. R. (1911). *Proc. R. Soc.* **A85**, 285.

Wu, C. S., Hayward, R. W., Hoppes, D. D., and Hudson, R. P. (1957). *Phys. Rev.* **105**, 1413.

Yukawa, H. (1935). *Proc. Phys. Math. Soc. Japan* **17**, 48.

Zweig, G. (1964). CERN reports TH401-412.

Answers to numerical problems

1.1. $V \geqslant 2.5$ kV.

1.3. (a) $Xl = 2T/e = 2 \times 10^3$ V, e.g. $l = 2$ cm, $X = 10^5$ V m^{-1} or 1 keV for plates 1 cm apart. (b) For $l = 2r$, $lB = 2\sin(22.5°)(2mT)^{\frac{1}{2}}/e = 8 \times 10^{-5}$ T m, e.g. $l = 2$ cm, $B = 4 \times 10^{-3}$ T.

1.4. 10^4 s^{-1}.

1.5. $\lambda = 124$ nm, $T_{max} = 5.28$ eV, $v_{max} = 1.36 \times 10^6$ m s^{-1}.

1.6. 3.9 per cent.

2.1. 1.2×10^{10} times as great, if pure ^{210}Po. 2.8×10^6 (i.e. less) for ^{226}Ra.

2.2. Within 1.5×10^{-7} of complete recovery.

2.5. (a) 1.53×10^{-7} m^3 (0.153 ml); (b) 9.26×10^{-4} ml.

2.6. 6×10^9 years.

2.8. (a) $R_p = 4.3$ cm, $R_d = R_\alpha = 6.12$ cm; (b) $R_p = R_d = R_\alpha = 36$ cm.

2.10. $T/m_0 c^2 = \beta^2/2 = 0.06593$ (classical), $T/m_0 c^2 = (1 - \beta^2)^{-\frac{1}{2}} - 1$ (relativistic) $= 0.0736$, hence $T_e = 37.4$ keV (relativistic).

3.1. $B = (2T_\alpha M_\alpha)^{1/2}/ed(2 + 2^{\frac{1}{2}})$, for angle of incidence 45°, $= 3.2 \times 10^5$ T.

3.2. $A = 135$.

3.3. $d\sigma/d\Omega = 4.12$ b sr^{-1}; $N_\theta = 0.027$ s^{-1}.

3.5. $f = A(Zn)Z^2(Cd)/A(Cd)Z^2(Zn)$, $Z(Cd) = 48$.

3.6. (a) 117 keV; (b) 1.7 eV.

4.1. 300 MeV.

4.2. 4.9 cm.

4.3. 0.9 mm.

4.4. 0.55.

4.5. (a) 1.46 m; (b) 0.54 T.

4.7. $\delta E = (3.5 \times 10^9$ MeV$)\,\delta t$. For $\delta t = \pm 0.4$ ns, $\delta E = \pm 1.4$ MeV.

4.9. $R^3 B^4 = 1395$; for $B = 1$ T, $R = 11.2$ m, for $B = 1.5$ T, $R = 6.5$ m.

5.1. (a) $T/2$; (b) R proportional to T/M.

5.2. $\Delta T = T_0\{1 - (A^2 + 2A\cos\varphi + 1)/(A + 1)^2\}$,
$$\cos\theta = \frac{A\cos\varphi + 1}{(A^2 + 2A\cos\varphi + 1)^{\frac{1}{2}}}.$$

5.3. $T_e = hv\{2\alpha/(1 + 2\alpha)\}$ for $\alpha = hv/m_e c^2$, 0.963, 1.119 MeV.

5.4. $d(Al)/d(Pb) = 13.13$, 162.4, 4.83, 13.54, 16.74.

5.6. $t = (1 - R_0\tau)^2/R_0^3\tau = 17.4$ s.

5.7. $\Delta V = 5.34$ mV.

5.8. $m_0 c^2 = 108$ MeV (i.e. a muon).

5.9. $6.3\,l$.

5.10. $m_0 c^2 = 940$ MeV, hence a proton.

5.11. 6202 years.

5.12. 6.13 Ci (2.27×10^{11} Bq), 36.5°C.

5.13. 111 min.

6.1. $hv = 2.22$ MeV, $T_d = 1.32$ keV.

6.2. $Q = T_1(m_1/m_4 - 1) + T_3(m_3/m_4 + 1) - 2\{(m_1/m_4)T_1 T_3\}^{1/2}$.

6.4. (a) $\simeq 10$ MeV; (b) 66.927150 u.

6.5. 1.14 MeV.

6.7. e-capture, $Q = 0.81$ MeV.

6.8. 2.0×10^4 b.

6.10. $\simeq 0.14$ g.

6.11. 256 MeV for $R = r_0 A^{\frac{1}{3}}$ with $r_0 = 1.2$ fm.

6.12. (a) 4×10^9 kg s^{-1}; (b) (i) 9.7×10^{-6}, (ii) 4.9×10^{-2}.

6.13. (a) 103 MeV; (b) 192 MeV, Maximum recoil energy $= 4T_0 m_1 m_2/ (m_1 + m_2)^2$. (Led to discovery of neutron—such observed recoils could not be explained by gamma-photons.)

7.3. $Kb = 1.831$, $V_0 = 33.8$ MeV, $V_0 b^2 = 1.49$ MeV b.

7.4. $d = 1.22$ fm.

7.5. Ratio is $0.70\ Z^2/0.72Z(Z-1)$.

7.6. $B = 1637$ MeV, $B/A = 7.87$ MeV, $B/m_p c^2 = 1.75$, (surface energy)$/B = -0.37$.

7.7. $Q_\alpha = 6.014, 4.843, 3.956, 3.221$ MeV.

7.8. $Z = (4a_4 + m_n - m_p)/(8a_4/A + 2a_3 A^{-1/3})$, $Z = 8,\ 19,\ 35$, in agreement with observations.

8.1. (a) 63 MeV, 23 MeV, mean $dE/d8 \simeq 67$ MeV cm^{-1}; (b) $T_p = 2.1$ MeV, range in air $= 7$ cm—but no visible change of curvature, and a 2 MeV proton would not pass through 6 mm of lead.

8.2. Relativistic: $183.5 - 129.6 = 53.9$ MeV; non-relativistic: $106.4 - 26.6 = 79.8$ MeV.

8.4. $T_\mu = 4.1$ MeV, $T_v = 29.8$ MeV, $T_\mu = (m_\pi c^2 - m_\mu c^2)^2/2m_\pi c^2$.

9.2. (a) 5/9, 2/3, 10/9, 11/9, 5/3 (b) 5/3, 2, 10/3, 11/3, 5.

9.3. p, Σ^+, Ξ^0, Σ^-, Ξ^-, n.

9.4. π^+, K$^+$, K^0, $\bar{\text{K}}^0$, π^-, K$^-$.

9.5. (a) 10, 20, 35, 56   (b) 9, 16, 25, 36.

Physical constants and conversion factors

Avogadro constant	L or N_A	6.022×10^{23} mol^{-1}
Bohr magneton	μ_B	9.274×10^{-24} J T^{-1}
Bohr radius	a_0	5.292×10^{-11} m
Boltzman constant	k	1.381×10^{-23} J K^{-1}
charge of an electron	e	-1.602×10^{-19} C
Compton wavelength of electron	$\lambda_C = h/m_e c = 2.426 \times 10^{-12}$ m	
Faraday constant	F	9.649×10^{4} C mol^{-1}
fine structure constant	$\alpha = \mu_0 e^2 c/2h = 7.297 \times 10^{-3}$ ($\alpha^{-1} = 137 \cdot 0$)	
gas constant	R	8.314 J K^{-1} mol^{-1}
gravitational constant	G	6.673×10^{-11} Nm^2 kg^{-2}
nuclear magneton	μ_N	5.051×10^{-27} J T^{-1}
permeability of a vacuum	μ_0	$4\pi \times 10^{-7}$ H m^{-1} exactly
permittivity of a vacuum	ε_0	8.854×10^{-12} F m^{-1} ($1/4\pi\varepsilon_0$ $= 8.988 \times 10^{9}$ m F^{-1})
Planck constant	h	6.626×10^{-34} J s
(Planck constant)/2π	$\hbar$	1.055×10^{-34} J s $= 6.582 \times 10^{-16}$ eV s
rest mass of electron	m_e	9.110×10^{-31} kg $= 0.511$ MeV/c^2
rest mass of proton	m_p	1.673×10^{-27} kg $= 938.3$ MeV/c^2
Rydberg constant	$R_\alpha = \mu_0^2 m_e e^4 c^3/8h^3 = 1.097 \times 10^{7}$ m^{-1}	
speed of light in a vacuum	c	2.998×10^{8} m s^{-1}
Stefan–Boltzmann constant	$\sigma = 2\pi^5 k^4/15h^3 c^2$ $= 5.670 \times 10^{-8}$ W m^{-2} K^{-4}	
unified atomic mass unit (^{12}C)	u	1.661×10^{-27} kg $= 931.5$ MeV/c^2
wavelength of a 1 eV photon		1.243×10^{-6} m

$1\,\text{Å} = 10^{-10}$ m; 1 dyne $= 10^{-5}$ N; 1 gauss (G) $= 10^{-4}$ tesla (T);
$0°\text{C} = 273.15$ K; 1 curie (Ci) $= 3.7 \times 10^{10}$ s^{-1};
1 J $= 10^7$ erg $= 6.241 \times 10^{18}$ eV; 1 eV $= 1.602 \times 10^{-19}$ J; 1 cal$_{th} = 4.184$ J;
$\ln 10 = 2.303$; $\ln x = 2.303 \log x$; e $= 2.718$; $\log$ e $= 0.4343$; $\pi = 3.142$

Index